Statistics for Environmental Science and Management

Bryan F.J. Manly

Statistical Consultant

Western EcoSystem Technology Inc.

Wyoming, USA

CHAPMAN & HALL/CRC

Boca Raton London New York Washington, D.C.

Library of Congress Cataloging-in-Publication Data

Manly, Bryan F.J., 1944–
Statistics for environmental science and management / by Bryan F.J. Manly.
p. cm.
Includes bibliographical references and index.
ISBN 1-58488-029-5 (alk. paper)
1. Environmental sciences — Statistical methods. 2. Environmental management — Statistical methods. I. Title.
GE45.S73 .M36 2000
363.7′007′27—dc21 00-055458
CIP

Direct all inquiries to CRC Press LLC, 2000 N.W. Corporate Blvd., Boca Raton, Florida 33431.

Visit the CRC Press Web site at www.crcpress.com

International Standard Book Number 1-58488-029-5
Library of Congress Card Number 00-055458
Printed in the United States of America 3 4 5 6 7 8 9 0
Printed on acid-free paper

A great deal of intelligence can be invested in ignorance when the need for illusion is deep.

Saul Bellow

Contents

Preface

This book is intended to introduce environmental scientists and managers to the statistical methods that will be useful for them in their work. A secondary aim was to produce a text suitable for a course in statistics for graduate students in the environmental science area. I wrote the book because it seemed to me that these groups should really learn about statistical methods in a special way. It is true that their needs are similar in many respects to those working in other areas. However, there are some special topics that are relevant to environmental science to the extent that they should be covered in an introductory text, although they would probably not be mentioned at all in such a text for a more general audience. I refer to environmental monitoring, impact assessment, assessing site reclamation, censored data, and Monte Carlo risk assessment, which all have their own chapters here.

The book is not intended to be a complete introduction to statistics. Rather, it is assumed that readers have already taken a course or read a book on basic methods, covering the ideas of random variation, statistical distributions, tests of significance, and confidence intervals. For those who have done this some time ago, Appendix A is meant to provide a quick refresher course.

A number of people have contributed directly or indirectly to this book. I must first mention Lyman McDonald of West Inc., Cheyenne, Wyoming, who first stimulated my interest in environmental statistics, as distinct from ecological statistics. Much of the contents of the book are influenced by the discussions that we have had on matters statistical. Jennifer Brown from the University of Canterbury in New Zealand has influenced the contents because we have shared the teaching of several short courses on statistics for environmental scientists and managers. Likewise, sharing a course on statistics for MSc students of environmental science with Caryn Thompson and David Fletcher has also had an effect on the book. Other people are too numerous to name, so I would just like to thank generally those who have contributed data sets, helped me check references and equations, etc.

Most of this book was written in the Department of Mathematics and Statistics at the University of Otago. As usual, the university was generous with the resources that are needed for the major effort of writing a book, including periods of sabbatical leave that enabled me to write large parts of the text without interruptions, and an excellent library.

However, the manuscript would definitely have taken longer to finish if I had not been invited to spend part of the year 2000 as a Visiting Researcher at the Max Planck Institute for Limnology at Plön in Germany. This enabled me to write the final chapters and put the whole book together. I am very grateful to Winfried Lampert, the Director of the Institute, for his kind invitation to come to Plön, and for allowing me to use the excellent facilities at the Institute while I was there.

The Saul Bellow quotation above may need some explanation. It results from attending meetings where an environmental matter is argued at length, with everyone being ignorant about the true facts of the case. Furthermore, one suspects that some people there would prefer not to know the true facts because this would be likely to end the arguments.

Bryan F.J. Manly
May 2000

CHAPTER 1

The Role of Statistics in Environmental Science

1.1 Introduction

In this chapter the role of statistics in environmental science is considered by examining some specific examples. First, however, an important point needs to be made. The importance of statistics is obvious because much of what is learned about the environment is based on numerical data. Therefore the appropriate handling of data is crucial. Indeed, the use of incorrect statistical methods may make individuals and organizations vulnerable to being sued for large amounts of money. Certainly in the United States it appears that increasing attention to the use of statistical methods is driven by the fear of litigation.

One thing that it is important to realize in this context is that there is usually not a single correct way to gather and analyse data. At best there may be several alternative approaches that are all about equally good. At worst the alternatives may involve different assumptions, and lead to different conclusions. This will become apparent from some of the examples in this and the following chapters.

1.2 Some Examples

The following examples demonstrate the non-trivial statistical problems that can arise in practice, and show very clearly the importance of the proper use of statistical theory. Some of these examples are revisited again in later chapters.

For environmental scientists and resource managers there are three broad types of situations that are often of interest:

(a) baseline studies intended to document the present state of the environment in order to establish future changes resulting, for example, from unforeseen events such as oil spills;

(b) targeted studies designed to assess the impact of planned events such as the construction of a dam, or accidents such as oil spills; and

(c) regular monitoring intended to detect trends and changes in important variables, possibly to ensure that compliance conditions are being met for an industry that is permitted to discharge small amounts of pollutants into the environment.

The examples include all of these types of situations.

Example 1.1 The Exxon Valdez Oil Spill

Oil spills resulting from the transport of crude and refined oils occur from time to time, particularly in coastal regions. Some very large spills (over 100,000 tonnes) have attracted considerable interest around the world. Notable examples are the *Torrey Canyon* spill in the English Channel in 1967, the *Amoco Cadiz* off the coast of Brittany, France in 1978, and the grounding of the *Braer* off the Shetland Islands in 1993. These spills all bring similar challenges for damage control for the physical environment and wildlife. There is intense concern from the public, resulting in political pressures on resource managers. There is the need to assess both short-term and long-term environmental impacts. Often there are lengthy legal cases to establish liability and compensation terms.

One of the most spectacular oil spills was that of the *Exxon Valdez*, which grounded on Bligh Reef in Prince William Sound, Alaska, on 24 March 1989, spilling more than 41 million litres of Alaska north slope crude oil. This was the largest spill up to that time in United States coastal waters, although far from the size of the *Amoco Cadiz* spill. The publicity surrounding it was enormous and the costs for cleanup, damage assessment and compensation have been considerable at nearly $US12,000 per barrel lost, compared with the more typical $US5,000 per barrel, for which the typical sale price is only about $US15 (Wells *et al.*, 1995, p. 5). Figure 1.1 shows the path of the oil through Prince William Sound and the western Gulf of Alaska.

There were many targeted studies of the *Exxon Valdez* spill related to the persistence and fate of the oil and the impact on fisheries and wildlife. Here only three of these studies, concerned with the shoreline impact of the oil, are considered. The investigators used different study designs and all met with complications that were not foreseen in advance of sampling. The three studies are Exxon's Shoreline Ecology Program (Page *et al.*, 1995; Gilfillan *et al.*, 1995), the Oil Spill Trustees' Coastal Habitat Injury Assessment (Highsmith *et al.*, 1993; McDonald *et al.*, 1995), and the Biological Monitoring Survey (Houghton *et al.*, 1993). The summary here owes much to a paper presented by Harner *et al.* (1995) at an International Environmetrics Conference in Kuala Lumpur, Malaysia.

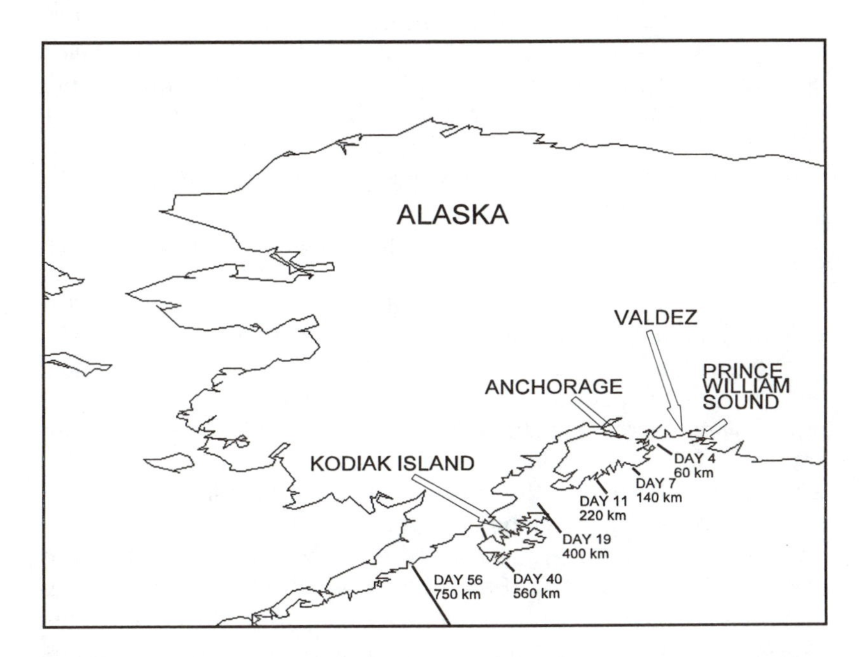

Figure 1.1 The path of the oil spill from the *Exxon Valdez* that occurred on 24 March (day 1) until 18 May 1989 (day 56), through Prince William Sound and the western Gulf of Alaska.

The Exxon Shoreline Ecology Program

The Exxon Shoreline Ecology Program started in 1989 with the purposeful selection of a number of heavily oiled sites along the shoreline that were to be measured over time in order to determine recovery rates. Because these sites are not representative of the shoreline potentially affected by oil they were not intended to assess the overall damage.

In 1990, using a stratified random sampling design of a type that is discussed in Chapter 2, the study was enlarged to include many more sites. Basically, the entire area of interest was divided into a number of short segments of shoreline. Each segment was then allocated to one of 16 strata based on the substrate type (exposed bedrock, sheltered bedrock, boulder/cobble, and pebble/gravel) and the degree of oiling (none, light, moderate, and heavy). For example, the first stratum was exposed bedrock with no oiling. Finally, four sites were chosen from each of the 16 strata for sampling to determine the abundances of more than a thousand species of animals and plants. A number of physical variables were also measured at each site.

The analysis of the data collected from the Exxon Shoreline Ecology Program was based on the use of what are called generalized linear models for species counts. These models are described in Chapter 3, and here it suffices to say that the effects of oiling were estimated on the assumption that the model used for each species was correct, with an allowance being made for differences in physical variables between sites.

A problem with the sampling design was that the initial allocation of shoreline segments to the 16 strata was based on the information in a geographical information system (GIS). However, this resulted in some sites being misclassified, particularly in terms of oiling levels. Furthermore, sites were not sampled if they were near an active eagle nest or human activity. The net result was that the sampling probabilities used in the study design were not quite what they were supposed to be. The investigators considered that the effect of this was minor. However, the authors of the National Oceanic and Atmospheric Administrations guidance document for assessing the damage from oil spills argue that this could be used in an attempt to discredit the entire study (Bergman *et al.*, 1995, Section F). It is therefore an example of how a minor deviation from the requirements of a standard study design may lead to potentially very serious consequences.

The Oil Spill Trustees' Coastal Habitat Injury Assessment

The *Exxon Valdez* Oil Spill Trustee Council was set up to oversee the allocation of funds from Exxon for the restoration of Prince William Sound and Alaskan waters. Like the Exxon Shoreline Ecology Program, the 1989 Coastal Habitat Injury Assessment study that was set up by the Council was based on a stratified random sampling design of a type that will be discussed in Chapter 3. There were 15 strata used, with these defined by five habitat types, each with three levels of oiling. Sample units were shoreline segments with varying lengths, and these were selected using a GIS system, with probabilities proportional to their lengths.

Unfortunately, so many sites were misclassified by the GIS system that the 1989 study design had to be abandoned in 1990. Instead, each of the moderately and heavily oiled sites that were sampled in 1989 was matched up with a comparable unoiled control site based on physical characteristics, to give a paired comparison design. The investigators then considered whether the paired sites were significantly different with regard to species abundance.

There are two aspects of the analysis of the data from this study that are unusual. First, the results of comparing site pairs (oiled and unoiled) were summarised as p-values (probabilities of observing differences as large as those seen on the hypothesis that oiling had no effect). These p-values were then combined using a meta-analysis which is a method for combining data that is described in Chapter 4. This method for assessing the evidence was used because each site pair was thought to be an independent study of the effects of oiling.

The second unusual aspect of the analysis was the weighting of results that was used for one of the two methods of meta-analysis that was employed. By weighting the results for each site pair by the reciprocal of the probability of the pair being included in the study, it was possible to make inferences with respect to the entire set of possible pairs in the study region. This was not a particularly simple procedure to carry out because inclusion probabilities had to be estimated by simulation. It did, however, overcome the problems introduce by the initial misclassification of sites.

The Biological Monitoring Survey

The Biological Monitoring Survey was instigated by the National Oceanic and Atmospheric Administration to study differences in impact between oiling alone and oiling combined with high pressure hot water washing at sheltered rocky sites. Thus there were three categories of sites used. Category 1 sites were unoiled. Category 2 sites were oiled but not washed. Category 3 sites were oiled and washed. Sites were subjectively selected, with unoiled ones being chosen to match those in the other two categories. Oiling levels were also classified as being light or moderate/heavy depending on their state when they were laid out in 1989. Species counts and percentage cover were measured at sampled sites.

Randomization tests were used to assess the significance of the differences between the sites in different categories because of the extreme nature of the distributions found for the recorded data. These types of test are discussed in Chapter 4. Here it is just noted that the hypothesis tested is that an observation was equally likely to have occurred for a site in any one of the three categories. These tests can certainly provide valid evidence of differences between the categories. However, the subjective methods used to select sites allow the argument to be made that any significant differences were due to the selection procedure rather than the oiling or the hot water treatment.

Another potential problem with the analysis of the study is that it may have involved pseudoreplication (treating correlated data as independent data), which is also defined and discussed in Chapter 4. This is because sampling stations along a transect on a beach were treated as if they provided completely independent data, although in fact some of these stations were in close proximity. In reality, observations taken close together in space can be expected to be more similar than observations taken far apart. Ignoring this fact may have led to a general tendency to conclude that sites in the different categories differed when this was not really the case.

General Comments on the Three Studies

The three studies on the *Exxon Valdez* oil spill took different approaches and lead to answers to different questions. The Exxon Shoreline Ecology Program was intended to assess the impact of oiling over the entire spill zone by using a stratified random sampling design. A minor problem is that the standard requirements of the sampling design were not quite followed because of site misclassificatio n and some restrictions on sites that could be sampled. The Oil Trustees' Coastal Habitat Study was badly upset by site misclassification in 1989, and was therefore converted to a paired comparison design in 1990 to compare moderately or heavily oiled sites with subjectively chosen unoiled sites. This allowed evidence for the effect of oiling to be assessed, but only at the expense of a complicated analysis involving the use of simulation to estimate the probability of a site being used in the study, and a special method to combine the results for different pairs of sites. The Biological Monitoring Survey focussed on assessing the effects of hot water washing, and the design gives no way for making inferences to the entire area affected by the oil spill.

All three studies are open to criticism in terms of the extent to which they can be used to draw conclusions about the overall impact of the oil spill in the entire area of interest. For the Exxon Coastal Ecology Program and the Trustees' Coastal Habitat Injury Assessment, this was the result of using stratified random sampling designs for which the randomization was upset to some extent. As a case study the *Exxon Valdez* oil spill should, therefore, be a warning to those involved in oil spill impact assessment in the future about problems that are likely to occur with this type of design. Another aspect of these two studies that should give pause for thought is that the analyses that had to be conducted were rather complicated and might have been difficult to defend in a court of law. They were not in tune with the KISS philosophy (**K**eep **I**t **S**imple **S**tatistician).

Example 1.2 Acid Rain in Norway

A Norwegian research programme was started in 1972 in response to widespread concern in Scandinavian countries about the effects of acid precipitation (Overrein *et al.*, 1980). As part of this study, regional surveys of small lakes were carried out in 1974 to 1978, with some extra sampling done in 1981. Data were recorded for pH, sulphate (SO_4) concentration, nitrate (NO_3) concentration, and calcium (Ca) concentration at each sampled lake. This can be considered a targeted study in terms of the three types of study that were defined in Section 1.1, but it may also be viewed as a monitoring study that was only continued for a relatively short period of time. Either way, the purpose of the study was to detect and describe changes in the water chemical variables that might be related to acid precipitation.

Table 1.1 shows the data from the study, as provided by Mohn and Volden (1985). Figure 1.2 shows the pH values, plotted against the locations of lakes in each of the years 1976, 1977, 1978 and 1981. Similar plots can, of course, be produced for sulphate, nitrate and calcium. The lakes that were measured varied from year to year. There is therefore a problem with missing data for some analyses that might be considered.

In practical terms, the main questions that are of interest from this study are:

(a) Is there any evidence of trends or abrupt changes in the values for one or more of the four measured chemistry variables?

(b) If trends or changes exist, are they related for the four variables, and are they of the type that can be expected to result from acid precipitation?

Table 1.1 Values for pH, sulphate (SO_4) concentration, nitrate (NO_3) concentration, and calcium (Ca) concentration for lakes in southern Norway with the latitudes (Lat) and longitudes (Long) for the lakes. Concentrations are in milligrams per litre. The sampled lakes varied to some extent from year to year because of the expense of sampling

			pH				SO_4				NO_3				CA			
Lake	Lat	Long	1976	1977	1978	1981	1976	1977	1978	1981	1976	1977	1978	1981	1976	1977	1978	1981
1	58.0	7.2	4.59		4.48	4.63	6.5		7.3	6.0	320		420	340	1.32		1.21	1.08
2	58.1	6.3	4.97		4.60	4.96	5.5		6.2	4.8	160		335	185	1.32		1.02	1.04
4	58.5	7.9	4.32	4.23	4.40	4.49	4.8	6.5	4.6	3.6	290	570	295	220	0.52	0.62	0.55	0.47
5	58.6	8.9	4.97	4.74	4.98	5.21	7.4	7.6	6.8	5.6	290	410	180	120	2.03	1.95	1.95	1.64
6	58.7	7.6	4.58	4.55	4.57	4.69	3.7	4.2	3.3	2.9	160	390	200	110	0.66	0.52	0.44	0.51
7	59.1	6.5	4.80		4.74	4.94	1.8		1.5	1.8	140		155	140	0.26		0.40	0.23
8	58.9	7.3	4.72	4.81	4.83	4.90	2.7	2.7	2.3	2.1	180	170	60	70	0.59	0.50	0.43	0.39
9	59.1	8.5	4.53	4.70	4.64	4.54	3.8	3.7	3.6	3.8	170	120	170	200	0.51	0.46	0.49	0.45
10	58.9	9.3	4.96	5.35	5.54	5.75	8.4	9.1	8.8	8.7	380	590	350	370	2.22	2.88	2.67	2.52
11	59.4	6.4	5.31	5.14	4.91	5.43	1.6	2.6	1.8	1.5	50	100	60	50	0.53	0.66	0.47	0.67
12	58.8	7.5	5.42	5.15	5.23	5.19	2.5	2.7	2.8	2.9	320	130	130	160	0.69	0.62	0.66	0.66
13	59.3	7.6	5.72		5.73	5.70	3.2		2.7	2.9	90		30	40	1.43		1.35	1.21
15	59.3	9.8	5.47		5.38	5.38	4.6		4.9	4.9	140		145	160	1.54		1.67	1.39
17	59.1	11.8	4.87	4.76	4.87	4.90	7.6	9.1	9.6	7.6	130	130	125	120	2.22	2.28	2.30	1.87
18	59.7	6.2	5.87	5.95	5.59	6.02	1.6	2.4	2.6	2.0	90	120	185	60	0.78	1.04	1.05	0.78
19	59.7	7.3	6.27	6.28	6.17	6.25	1.5	1.3	1.9	1.7	10	20	15	10	1.15	0.97	1.14	1.04
20	59.9	8.3	6.67	6.44	6.28	6.67	1.4	1.6	1.8	1.8	20	30	10	10	2.47	1.14	1.18	2.34
21	59.8	8.9	6.06		5.80	6.09	4.6		5.3	4.2	30		20	50	2.18		2.08	1.99
24	60.1	12.0	5.38	5.32	5.33	5.21	5.8	6.2	5.9	5.4	50	130	45	50	2.10	2.20	1.94	1.79
26	59.6	5.9	5.41	5.94			1.5	1.6			220	90			0.61	0.65		
30	60.4	10.2	5.60	6.10	5.57	5.98	4.0	3.9	4.9	4.3	30	50	165	60	1.86	2.24	2.25	2.18

Table 1.1

Lake	Lat	Long	pH				SO_4				NO_3				CA			
			1976	1977	1978	1981	1976	1977	1978	1981	1976	1977	1978	1981	1976	1977	1978	1981
32	60.4	12.2	4.93	4.94	4.91	4.93	5.1	5.7	5.4	4.3	70	110	80	70	1.45	1.56	1.44	1.26
34-1	60.5	5.5			4.90	4.87			1.4	1.3			175	90			0.37	0.19
36	60.9	7.3	5.60	5.69	5.41	5.66	1.4	1.0	1.1	1.2	70	70	60	70	0.46	0.34	0.74	0.37
38	60.9	10.0	6.72	6.59	6.39		3.8	3.3	3.1		30	30	20		2.67	2.53	2.50	
40	60.7	12.2	5.97	6.02	5.71	5.67	5.1	5.8	5.0	4.2	60	130	50	50	2.19	2.28	2.06	1.85
41	61.0	5.0	4.68	4.72	5.02		2.8	3.2	1.6		70	160	50		0.47	0.48	0.34	
42	61.3	5.6	5.07			5.18	1.6			1.6	40			30	0.49			0.37
43	61.0	6.9	6.23	6.34	6.20	6.29	1.5	1.5	1.4	1.6	50	60	20	40	1.56	1.53	1.68	1.54
46	61.0	9.7	6.64		6.24	6.37	3.2		2.6	2.3	70		30	50	2.49		2.14	2.07
47	61.3	10.8	6.15	6.23	6.07	5.68	2.8	1.7	1.9	1.8	100	30	15	200	2.00	0.96	2.04	2.68
49	61.5	4.9	4.82	4.77	5.09	5.45	3.0	1.9	1.5	1.7	100	150	100	100	0.44	0.36	0.41	0.32
50	61.5	5.5	5.42	4.82	5.34	5.54	0.7	1.8	1.5	1.5	40	360	60	50	0.32	0.55	0.58	0.48
57	61.7	4.9	4.99		5.16	5.25	3.1		2.4	2.2	30		20	10	0.84		0.91	0.53
58	61.7	5.8	5.31	5.77	5.60	5.55	2.1	1.9	1.3	1.6	20	90	20	10	0.69	0.57	0.66	0.64
59	61.9	7.1	6.26	5.03	5.85		3.9	1.5	1.7		70	240	20		2.24	0.58	0.73	
65	62.2	6.4	5.99	6.10	5.99	6.13	1.9	1.9	1.5	1.7	10	40	10	10	0.69	0.76	0.80	0.66
80	58.1	6.7	4.63		4.59	4.92	5.2		5.6	3.9	290		315	85	0.85		0.81	0.77
81	58.3	8.0	4.47		4.36	4.50	5.3		5.4	4.2	250		425	100	0.87		0.82	0.55
82	58.7	7.1	4.60		4.54	4.66	2.9		2.9	2.2	150		110	60	0.61		0.65	0.48
83	58.9	6.1	4.88	4.99	4.86	4.92	1.6	1.5	1.7	1.9	140	130	165	130	0.36	0.22	0.33	0.25
85	59.4	11.3	4.60	4.88	4.91	4.84	13.0	15.0	13.0	10.0	380	90	180	280	3.47	3.72	3.05	2.61
86	59.3	9.4	4.85	4.65	4.77	4.84	5.5	5.9	5.7	4.8	90	140	150	160	1.70	1.65	1.65	1.30
87	59.2	7.6	5.06		5.15	5.11	2.8		2.6	3.0	90		70	120	0.81		0.84	0.73
88	59.4	7.3	5.97	5.82	5.90	6.17	1.6	1.6	1.4	1.8	60	190	65	40	0.83	0.91	0.96	0.89
89	59.3	6.3	5.47		6.05	5.82	2.0		2.4	2.0	110		95	10	0.79		1.22	0.76
94	61.0	11.5	6.05	5.97	5.78	5.75	5.8	6.9	5.9	5.8	50	100	70	50	2.91	2.79	2.64	1.24
95-1	61.2	4.6			5.70	5.50			2.3	1.6			240	70			0.94	0.59
		Mean	5.34	5.40	5.31	5.38	3.74	3.98	3.72	3.33	124.1	161.6	124.1	100.2	1.29	1.27	1.23	1.08
		SD	0.65	0.66	0.57	0.56	2.32	3.06	2.53	2.03	101.4	144.0	110.1	83.9	0.81	0.90	0.74	0.71

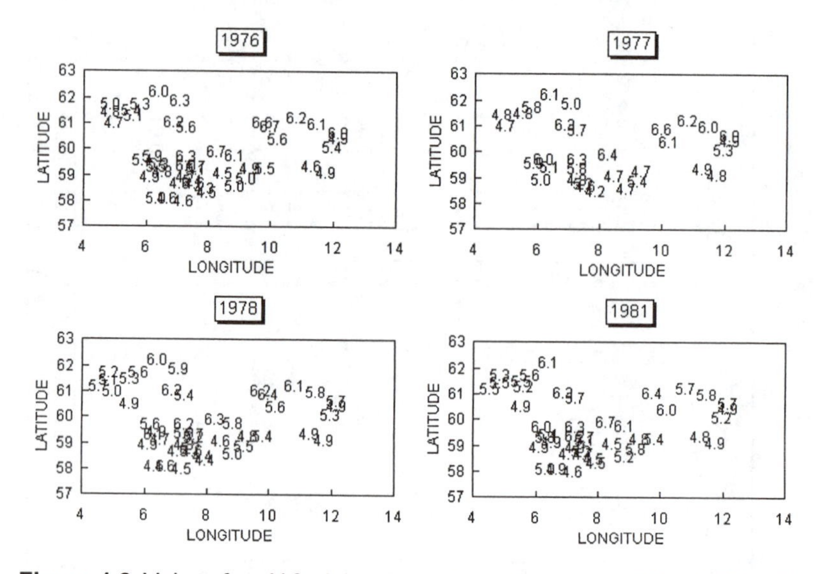

Figure 1.2 Values for pH for lakes in southern Norway in 1976, 1977, 1978 and 1981, plotted against the longitude and latitude of the lakes.

Other questions that may have intrinsic interest but are also relevant to the answering of the first two questions are:

(c) Is there evidence of spatial correlation such that measurements on lakes that are in close proximity tend to be similar?

(d) Is there evidence of time correlation such that the measurements on a lake tend to be similar if they are close in time?

One of the important considerations in many environmental studies is the need to allow for correlation in time and space. Methods for doing this are discussed at some length in Chapters 8 and 9, as well as being mentioned briefly in several other chapters. Here it can merely be noted that a study of the pH values in Figure 1.2 indicates a tendency for the highest values to be in the north, with no striking changes from year to year for individual lakes (which are, of course, plotted at the same location for each of the years they were sampled).

Example 1.3 Salmon Survival in the Snake River

The Snake River and the Columbia River in the Pacific northwest of the United States contain eight dams used for the generation of electricity, as shown in Figure 1.3. These rivers are also the migration route for hatchery and wild salmon, so there is a clear potential for conflict between different uses of the rivers. The dams were constructed with bypass systems for the salmon, but there has been concern nevertheless about salmon mortality rates in passing downstream, with some studies suggesting losses as high as 85% of hatchery fish in just portions of the river.

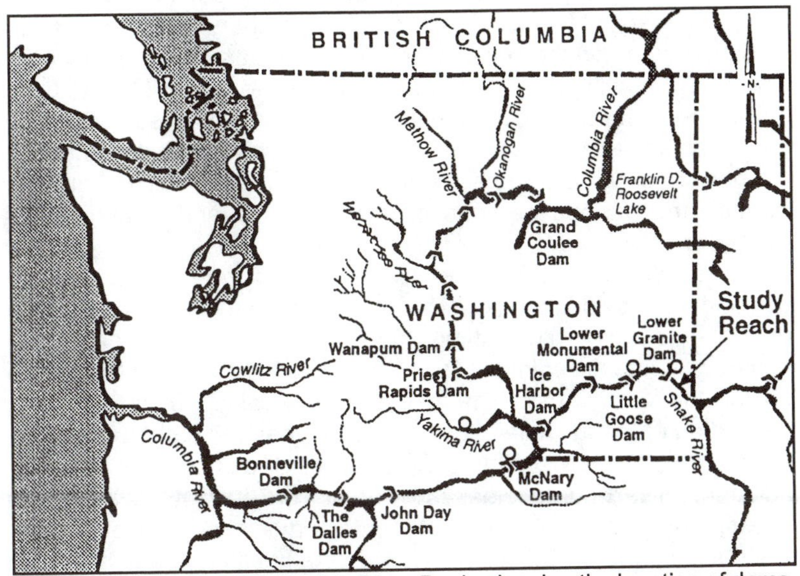

Figure 1.3 Map of the Columbia River Basin showing the location of dams. Primary releases of pit-tagged salmon were made in 1993 and 1994 above Lower Granite Dam, with recoveries at Lower Granite Dam and Little Goose Dam in 1993, and at these dams plus Lower Monumental Dam in 1994.

In order to get a better understanding of the causes of salmon mortality, a major study was started in 1993 by the National Marine Fisheries Service and the University of Washington to investigate the use of modern mark-recapture methods for estimating survival rates through both the entire river system and the component dams. The methodology was based on theory developed by Burnham *et al.* (1987) specifically for mark-recapture experiments for estimating the survival of fish through dams, but with modifications designed for the application in question (Dauble *et al.*, 1993). Fish are fitted with Passive Integrated Transponder (PIT) tags which can be uniquely identified at downstream detection stations in the bypass systems of dams. Batches of tagged fish are released and their recoveries at detection stations are recorded. Using special probability models, it is then possible to use the recovery information to estimate the probability of a fish surviving through different stretches of the rivers and the probability of fish being detected as they pass through a dam.

In 1993 a pilot programme of releases were made to (a) field test the mark-recapture method for estimating survival, including testing the assumptions of the probability model; (b) identify operational and logistic constraints limiting the collection of data; and (c) determine whether survival estimates could be obtained with adequate precision. Seven primary batches of 830 to 1442 hatchery yearling chinook salmon (*Oncorhynchus tshawytscha*) were released above the Lower Granite Dam, with some secondary releases at Lower Granite Dam and Little Goose Dam to measure the mortality associated with particular aspects of the dam system. It was concluded that the methods used will provide accurate estimates of survival probabilities through the various sections of the Columbia and Snake Rivers (Iwamoto *et al.*, 1994).

The study continued in 1994 with ten primary releases of hatchery yearling chinook salmon (*O. tshawytscha*) in batches of 542 to 1196, one release of 512 wild yearling chinook salmon, and nine releases of hatchery steelhead salmon (*O. mykiss*) in batches of 1001 to 4009, all above the first dam. The releases took place over a greater proportion of the juvenile migration period than in 1993, and survival probabilities were estimated for a larger stretch of the river. In addition, 58 secondary releases in batches of 700 to 4643 were made to estimate the mortality associated with particular aspects of the dam system. In total, the records for nearly 100,000 fish were analysed so that this must be one of the largest mark-recapture study ever carried out in one year with uniquely tagged individuals. From the results obtained the researchers concluded that the assumptions of the models used were generally satisfied and reiterated their belief that these models permit the accurate estimation of survival probabilities through individual river sections, reservoirs and dams on the Snake and Columbia Rivers (Muir *et al.*, 1995).

In terms of the three types of study that were defined in Section 1.1, the mark-recapture experiments on the Snake River in 1993 and 1994 can be thought of as part of a baseline study because the main objective was to assess this approach for estimating survival rates of salmon with the present dam structures with a view to assessing the value of possible modifications in the future. Estimating survival rates for populations living outside captivity is usually a difficult task, and this is certainly the case for salmon in the Snake and Columbia Rivers. However, the estimates obtained by mark-recapture seem quite accurate, as is indicated by the results shown in Table 1.2.

Table 1.2 Estimates of survival probabilities for ten releases of hatchery yearling chinook salmon made above the Lower Granite Dam in 1994 (Muir *et al.*, 1995). The survival is through the Lower Granite Dam, Little Goose Dam and Lower Monumental Dam. The standard errors shown with individual estimates are calculated from the mark-recapture model. The standard error of the mean is the standard deviation of the ten estimates divided by $\sqrt{10}$

Release Date	Number Released	Survival Estimate	Standard Error
16-Apr	1189	0.688	0.027
17-Apr	1196	0.666	0.028
18-Apr	1194	0.634	0.027
21-Apr	1190	0.690	0.040
23-Apr	776	0.606	0.047
26-Apr	1032	0.630	0.048
29-Apr	643	0.623	0.069
1-May	1069	0.676	0.056
4-May	542	0.665	0.094
10-May	1048	0.721	0.101
	Mean	0.660	0.011

Future objectives of the research programme include getting a good estimate of the survival rate of salmon for a whole migration season for different parts of the river system, allowing for the possibility of time changes and trends. These objectives pose interesting design problems, with the need to combine mark-recapture models with more traditional finite sampling theory, as discussed in Chapter 2.

This example is unusual because of the use of the special mark-recapture methods. It is included here to illustrate the wide variety of statistical methods that are applicable for solving environmental problems – in this case improving the survival of salmon in a river that is used for electricity generation.

Example 1.4 A Large-Scale Perturbation Experiment

Predicting the responses of whole ecosystems to perturbations is one of the greatest challenges to ecologists because this often requires experimental manipulations to be made on a very large-scale. In many cases small-scale laboratory or field experiments will simply not necessarily demonstrate the responses obtained in the real world. For this reason a number of experiments have been conducted on lakes, catchments, streams, and open terrestrial and marine environments. Although these experiments involve little or no replication, they do indicate the response potential of ecosystems to powerful manipulations which can be expected to produce massive unequivocal changes (Carpenter *et al.*, 1995). They are targeted studies as defined in Section 1.1.

Carpenter *et al.* (1989) discussed some examples of large-scale experiments involving lakes in the Northern Highlands Lake District of Wisconsin in the United States. One such experiment, which was part of the Cascading Trophic Interaction Project, involved removing 90% of the piscivore biomass from Peter Lake and adding 90% of the planktivore biomass from another lake. Changes in Peter Lake over the following two years were then compared with changes in Paul Lake, which is in the same area but received no manipulation. Studies of this type are often referred to as having a before-after-control-impact (BACI) design, of a type that is discussed in Chapter 6.

One of the variables measured at Peter Lake and Paul Lake was the chlorophyll concentration in mg/m^3. This was measured for ten samples taken in June to August 1984, for 17 samples taken in June to August 1985, and for 15 samples taken in June to August 1986. The manipulation of Peter Lake was carried out in May 1985. Figure 1.4 shows the results obtained. In situations like this the hope is that time effects other than those due to the manipulation are removed by taking the difference between measurements for the two lakes. If this is correct, then a comparison between the mean difference between the lakes before the manipulation with the mean difference after the manipulation gives a test for an effect of the manipulation.

Before the manipulation, the sample size is 10 and the mean difference (treated - control) is -2.020. After the manipulation the sample size is 32 and the mean difference is -0.953. To assess whether the change in the mean difference is significant, Carpenter *et al.* (1989) used a randomization test. This involved comparing the observed change with the distribution obtained for this statistic by randomly reordering the time series of differences, as discussed further in Section 4.6. The outcome of this test was significant at the 5% level so they concluded that there was evidence of a change.

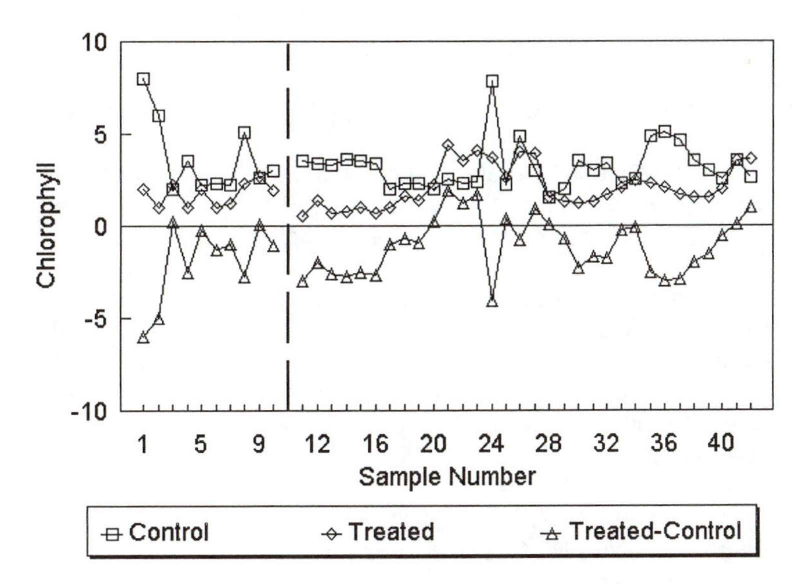

Figure 1.4 The outcome of an intervention experiment in terms of chlorophyll concentrations (mg/m^3). Samples 1 to 10 were taken in June to August 1984, samples 11 to 27 were taken from June to August 1985, and samples 28 to 42 were taken in June to August 1986. The treated lake received a food web manipulation in May 1985, between sample numbers 10 and 11 (as indicated by a broken vertical line).

A number of other statistical tests to compare the mean differences before and after the change could have been used just as well as the randomization test. However, most of these tests may be upset to some extent by correlation between the successive observations in the time series of differences between the manipulated and the control lake. Because this correlation will generally be positive it has the tendency to give more significant results than should otherwise occur. From the results of a simulation study, Carpenter *et al.* (1989) suggested that this can be allowed for by regarding effects that are significant between the 1 and 5% level as equivocal if correlation seems to be present. From this point of view the effect of the manipulation of Peter Lake on the chlorophyll concentration is not clearly established by the randomization test.

This example demonstrates the usual problems with BACI studies. In particular:

(a) the assumption that the distribution of the difference between Peter Lake and Paul Lake would not have changed with time in the absence of any manipulation is not testable, and making this assumption amounts to an act of faith; and

(b) the correlation between observations taken with little time between them is likely to be only partially removed by taking the difference between the results for the manipulated lake and the control lake, with the result that the randomization test (or any simple alternative test) for a manipulation effect is not completely valid.

There is nothing that can be done about problem (a) because of the nature of the situation. More complex time series modelling offers the possibility of overcoming problem (b), but there are severe difficulties with using these techniques with the relatively small sets of data that are often available. These matters are considered further in Chapters 6 and 8.

Example 1.5 Ring Widths of Andean Alders

Tree ring width measurements are useful indicators of the effects of pollution, climate, and other environmental variables (Fritts, 1976; Norton and Ogden, 1987). There is therefore interest in monitoring the widths at particular sites to see whether changes are taking place in the distribution of widths. In particular, trends in the distribution may be sensitive indicators of environmental changes.

With this in mind, Dr Alfredo Grau collected data on ring widths for 27 Andean alders (*Alnus acuminanta*) on the Taficillo Ridge at an altitude of about 1700 m in Tucuman, Argentina, every year from 1970 to 1989. The measurements that he obtained are shown in Figure 1.5. It is apparent here that over the period of the study the mean width decreased, as did the amount of variation between individual trees. Possible reasons for a change of the type observed here are climate changes and pollution. The point is that regularly monitored environmental indicators such as tree ring widths can be used to signal changes in conditions. The causes of these changes can then be investigated in targeted studies.

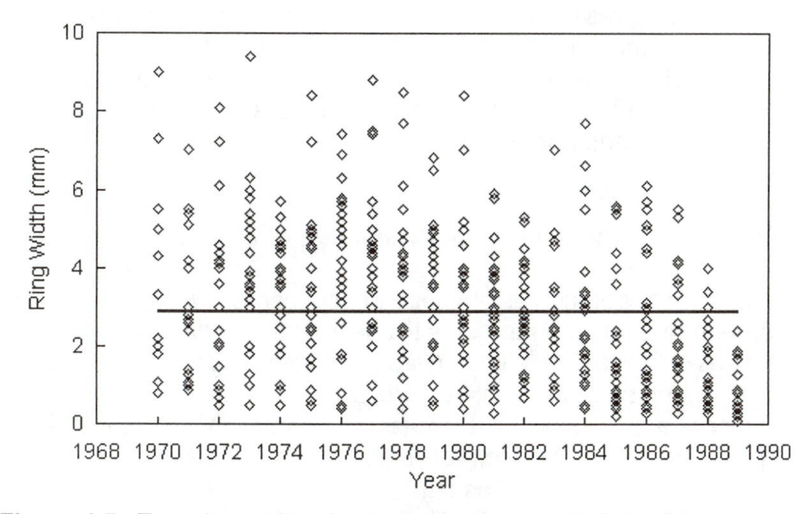

Figure 1.5 Tree ring widths for Andean alders on Taficillo Ridge, near Tucuman, Argentina, 1970-1989. The horizontal line is the overall mean for all ring widths in all years.

Example 1.6 Monitoring Antarctic Marine Life

An example of monitoring on a very large-scale is provided by work carried out by the Commission for the Conservation of Antarctic Marine Living Resources (CCAMLR), an intergovernmental organization established to develop measures for the conservation of marine life of the Southern Ocean surrounding Antarctica. Currently 21 countries are members of the Commission, while seven other states have acceded to the Convention set up as part of CCAMLR to govern the use of the resources in question (CCAMLR, 1992).

One of the working groups of CCAMLR is responsible for Ecosystem Monitoring and Management. Monitoring in this context involves the collection of data on indicators of the biological health of Antarctica. These indicators are annual figures that are largely determined by what is available as a result of scientific research carried out by member states. At present they include such things as the average weight of penguins when they arrive at various breeding colonies, the average time that penguins spend on the first shift incubating eggs, the catch of krill by fishing vessels within 100km of land-based penguin breeding sites, average foraging durations of fur seal cows, and the percentage cover of sea-ice. There are plans to considerably increase the number of indicators to include other species and more physical variables. Major challenges include ensuring that research groups of different nationalities collect data using the same standard methods and, in the longer term, being able to understand the relationships between different indicators and combining them better to measure the state of the Antarctic and detect trends and abrupt changes.

Example 1.7 Evaluating the Attainment of Cleanup Standards

Many environmental studies are concerned with the specific problem of evaluating the effectiveness of the reclamation of a site that has suffered from some environmental damage. For example, a government agency might require a mining company to work on restoring a site until the biomass of vegetation per unit area is equivalent to what is found on undamaged reference areas. This requires a targeted study as defined in Section 1.1.

There are two complications with using standard statistical methods in this situation. The first is that the damaged and reference sites are not generally selected randomly from populations of potential sites, and it is unreasonable to suppose that they would have had exactly the same mean for the study variable even in the absence of any impact on the damaged site. Therefore, if large samples are taken from each site there will be a high probability of detecting a difference, irrespective of the extent to which the damaged site has been reclaimed. The second complication is that when a test for a difference between the two sites does not give a significant result this does not necessarily mean that a difference does not exist. An alternative explanation is that the sample sizes were not large enough to detect a difference that does exist.

These complications with statistical tests have led to a recommendation by the United States Environmental Protection Agency (1989a) that the null hypothesis for statistical tests should depend on the status of a site, in the following way:

(a) If a site has not been declared to be contaminated, then the null hypothesis should be that it is clean, i.e., there is no difference from the control site. The alternative hypothesis is that the site is contaminated. A non-significant test result leads to the conclusion that there is no real evidence that the site is contaminated.

(b) If a site has been declared to be contaminated, then the null hypothesis is that this is true, i.e., there is a difference (in an unacceptable direction) from the control site. The alternative hypothesis is that the site is clean. A non-significant test result leads to the conclusion that there is no real evidence that the site has been cleaned up.

The point here is that once a site has been declared to have a certain status pertinent evidence should be required to justify changing this status.

If the point of view expressed by (a) and (b) is not adopted, so that the null hypothesis is always that the damaged site is not different from the control, then the agency charged with ensuring that the site is cleaned up is faced with setting up a maze of regulations to ensure that study designs have large enough sample sizes to detect differences of practical importance between the damaged and control sites. If this is not done, then it is apparent that any organization wanting to have the status of a site changed from contaminated to clean should carry out the smallest study possible, with low power to detect even a large difference from the control site. The probability of a non-significant test result (the site is clean) will then be as high as possible.

As an example of the type of data that may be involved in the comparison of a control site and a possibly contaminated one, consider some measurements of 1,2,3,4-tetrachlorobenzene (TcCB) in parts per thousand million given by Gilbert and Simpson (1992, p. 6.22). There are 47 measurements made in different parts of the control site and 77 measurements made in different parts of the possibly contaminated site, as shown in Table 1.3 and Figure 1.6. Clearly the TcCB levels are much more variable at the possibly contaminated site. Presumably this might have occurred from the TcCB levels being lowered in parts of the site by cleaning, while very high levels remained in other parts of the site.

Table 1.3 Measurements of 1,2,3,4-tetrachlorobenzene from samples taken at a reference site and a possibly contaminated site

Reference site (n = 47)											
0.60	0.50	0.39	0.84	0.46	0.39	0.62	0.67	0.69	0.81	0.38	0.79
0.43	0.57	0.74	0.27	0.51	0.35	0.28	0.45	0.42	1.14	0.23	0.72
0.63	0.50	0.29	0.82	0.54	1.13	0.56	1.33	0.56	1.11	0.57	0.89
0.28	1.20	0.76	0.26	0.34	0.52	0.42	0.22	0.33	1.14	0.48	
								Mean =	0.60	SD =	0.28

Possibly contaminated site (n = 75)											
1.33	0.09	0.12	0.28	0.14	0.16	0.17	0.47	0.17	0.18	0.19	0.09
18.40	0.20	0.21	0.12	0.22	0.22	0.22	168.6	0.24	0.25	0.25	0.20
0.48	0.26	5.56	0.21	0.29	0.31	0.33	3.29	0.33	0.34	0.37	0.25
2.59	0.39	0.40	0.28	0.43	6.61	0.48	0.17	0.49	0.51	0.51	0.38
0.92	0.60	0.61	0.43	0.75	0.82	0.85	0.23	0.94	1.05	1.10	0.54
1.53	1.19	1.22	0.62	1.39	1.39	1.52	0.33	1.73	2.35	2.46	1.10
51.97	2.61	3.06									
								Mean =	4.02	SD =	20.27

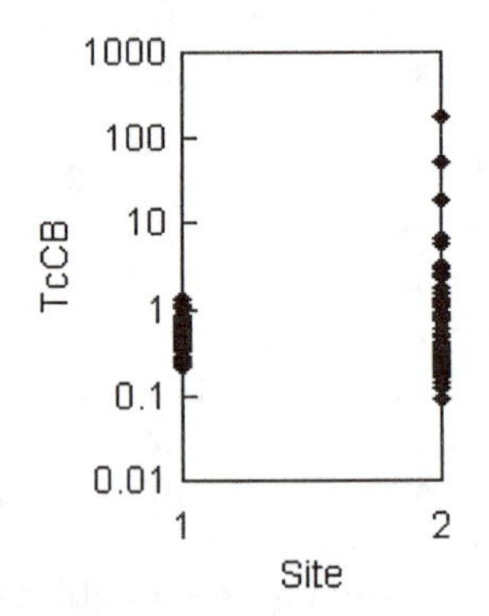

Figure 1.6 Comparison of TcCB measurements in parts per thousand million at a contaminated site (2) and a reference site (1).

Methods for comparing samples such as these in terms of means and variation are discussed further in Chapter 7. For the data in this example, the extremely skewed distribution at the contaminated site, with several very extreme values, should lead to some caution in making comparisons based on the assumption that distributions are normal within sites.

1.3 The Importance of Statistics in the Examples

The examples that are presented above demonstrate clearly the importance of statistical methods in environmental studies. With the *Exxon Valdez* oil spill, problems with the application of the study designs meant that rather complicated analyses were required to make inferences. With the Norwegian study on acid rain there is a need to consider the impact of correlation in time and space in the water quality variables that were measured. The estimation of the yearly survival rates of salmon in the Snake River requires the use of special models for analysing mark-recapture experiments combined with the use of the theory of sampling for finite populations. Monitoring studies such as the one involving the measurement of tree ring width in Argentina call for the use of methods for the detection of trends and abrupt changes in distributions. Monitoring of whole ecosystems as carried out by the Commission for the Conservation of Antarctic Marine Living Resources requires the collection and analysis of vast amounts of data, with many very complicated statistical problems. The comparison of samples from contaminated and reference sites may require the use of tests that are valid with extremely non-normal distributions. All of these matters are considered in some detail in the pages that follow.

1.4 Chapter Summary

- Statistics is important in environmental science because much of what is known about the environment comes from numerical data.

- Three broad types of study of interest to resource managers are baseline studies (to document the present state of the environment), targeted studies (to assess the impact of particular events), and regular monitoring (to detect trends and other changes in important variables).

- All types of study involve sampling over time and space and it is important that sampling designs are cost effective and can be justified in a court of law if necessary.

- Seven examples are discussed to demonstrate the importance of statistical methods to environmental science. These examples involve the shoreline impact of the *Exxon Valdez* oil spill in Prince William Sound, Alaska in March 1989; a Norwegian study of the possible impact of acid precipitation on small lakes; estimation of the survival rate of salmon in the Snake and Columbia Rivers in the Pacific Northwest of the United States; a large-scale perturbation experiment carried out in Wisconsin in the United States involving changing the piscivore and planktivore composition of a lake and comparing changes in the chlorophyll composition with changes in a control lake; monitoring of the annual ring widths of Andean alders near Tucuman in Argentina; monitoring marine life in the Antarctic, and comparing a possibly contaminated site with a control site in the United States, in terms of measurements of the amounts of a pollutant in samples taken from the two sites.

CHAPTER 2

Environmental Sampling

2.1 Introduction

All of the examples considered in the previous chapter involved sampling of some sort, showing that the design of sampling schemes is an important topic in environmental statistics. This chapter is therefore devoted to considering this topic in some detail. The estimation of mean values, totals and proportions from the data collected by sampling is conveniently covered at the same time, and this means that the chapter includes all that is needed for many environmental problems.

The first task in designing a sampling scheme is to define the population of interest, and the sample units that make up this population. Here the 'population' is defined as a collection of items that are of interest, and the 'sample units' are these items. In this chapter it is assumed that each of the items is characterised by the measurements that it has for certain variables (e.g., weight or height), or which of several categories it falls into (e.g., the colour that it possesses, or the type of habitat where it is found). When this is the case, statistical theory can assist in the process of drawing conclusions about the population using information from a sample of some of the items.

Sometimes defining the population of interest and the sample units is straightforward because the extent of the population is obvious, and a natural sample unit exists. However, at other times some more or less arbitrary definitions will be required. An example of a straightforward situation is where the population is all the farms in a region of a country and the variable of interest is the amount of water used for irrigation on a farm. This contrasts with the situation where there is interest in the impact of an oil spill on the flora and fauna on beaches. In that case the extent of the area that might be affected may not be clear, and it may not be obvious which length of beach to use as a sample unit. The investigator must then subjectively choose the potentially affected area, and impose a structure in terms of sample units. Furthermore, there will not be a 'correct' size for the sample unit. A range of lengths of beach may serve equally well, taking into account the method that is used to take measurements.

The choice of what to measure will also, of course, introduce some further subjective decisions.

2.2 Simple Random Sampling

A simple random sample is one that is obtained by a process that gives each sample unit the same probability of being chosen. Usually it will be desirable to choose such a sample without replacement so that sample units are not used more than once. This gives slightly more accurate results than sampling with replacement whereby individual units can appear two or more times in the sample. However, for samples that are small in comparison with the population size, the difference in the accuracy obtained is not great.

Obtaining a simple random sample is easiest when a sampling frame is available, where this is just a list of all the units in the population from which the sample is to be drawn. If the sampling frame contains units numbered from 1 to N, then a simple random sample of size n is obtained without replacement by drawing n numbers one by one in such a way that each choice is equally likely to be any of the numbers that have not already been used. For sampling with replacement, each of the numbers 1 to N is given the same chance of appearing at each draw.

The process of selecting the units to use in a sample is sometimes facilitated by using a table of random numbers such as the one shown in Table 2.1. As an example of how such a table can be used, suppose that a study area is divided into 116 quadrats as shown in Figure 2.1, and it is desirable to select a simple random sample of ten of these quadrats without replacement. To do this, first start at an arbitrary place in the table such as the beginning of row five. The first three digits in each block of five digits can then be considered, to give the series 698, 419, 008, 127, 106, 605, 843, 378, 462, 953, 745, and so on. The first ten different numbers between 1 and 116 then give a simple random sample of quadrats: 8, 106, and so on. For selecting large samples essentially the same process can be carried out on a computer using pseudo-random numbers in a spreadsheet, for example.

Table 2.1 A random number table with each digit chosen such that 0, 1, ..., 9 were equally likely to occur. The grouping into groups of four digits is arbitrary so that, for example, to select numbers from 0 to 99999 the digits can be considered five at a time

1252	9045	1286	2235	6289	5542	2965	1219	7088	1533
9135	3824	8483	1617	0990	4547	9454	9266	9223	9662
8377	5968	0088	9813	4019	1597	2294	8177	5720	8526
3789	9509	1107	7492	7178	7485	6866	0353	8133	7247
6988	4191	0083	1273	1061	6058	8433	3782	4627	9535
7458	7394	0804	6410	7771	9514	1689	2248	7654	1608
2136	8184	0033	1742	9116	6480	4081	6121	9399	2601
5693	3627	8980	2877	6078	0993	6817	7790	4589	8833
1813	0018	9270	2802	2245	8313	7113	2074	1510	1802
9787	7735	0752	3671	2519	1063	5471	7114	3477	7203
7379	6355	4738	8695	6987	9312	5261	3915	4060	5020
8763	8141	4588	0345	6854	4575	5940	1427	8757	5221
6605	3563	6829	2171	8121	5723	3901	0456	8691	9649
8154	6617	3825	2320	0476	4355	7690	9987	2757	3871
5855	0345	0029	6323	0493	8556	6810	7981	8007	3433
7172	6273	6400	7392	4880	2917	9748	6690	0147	6744
7780	3051	6052	6389	0957	7744	5265	7623	5189	0917
7289	8817	9973	7058	2621	7637	1791	1904	8467	0318
9133	5493	2280	9064	6427	2426	9685	3109	8222	0136
1035	4738	9748	6313	1589	0097	7292	6264	7563	2146
5482	8213	2366	1834	9971	2467	5843	1570	5818	4827
7947	2968	3840	9873	0330	1909	4348	4157	6470	5028
6426	2413	9559	2008	7485	0321	5106	0967	6471	5151
8382	7446	9142	2006	4643	8984	6677	8596	7477	3682
1948	6713	2204	9931	8202	9055	0820	6296	6570	0438
3250	5110	7397	3638	1794	2059	2771	4461	2018	4981
8445	1259	5679	4109	4010	2484	1495	3704	8936	1270
1933	6213	9774	1158	1659	6400	8525	6531	4712	6738
7368	9021	1251	3162	0646	2380	1446	2573	5018	1051
9772	1664	6687	4493	1932	6164	5882	0672	8492	1277
0868	9041	0735	1319	9096	6458	1659	1224	2968	9657
3658	6429	1186	0768	0484	1996	0338	4044	8415	1906
3117	6575	1925	6232	3495	4706	3533	7630	5570	9400
7572	1054	6902	2256	0003	2189	1569	1272	2592	0912
3526	1092	4235	0755	3173	1446	6311	3243	7053	7094
2597	8181	8560	6492	1451	1325	7247	1535	8773	0009
4666	0581	2433	9756	6818	1746	1273	1105	1919	0986
5905	5680	2503	0569	1642	3789	8234	4337	2705	6416
3890	0286	9414	9485	6629	4167	2517	9717	2582	8480
3891	5768	9601	3765	9627	6064	7097	2654	2456	3028

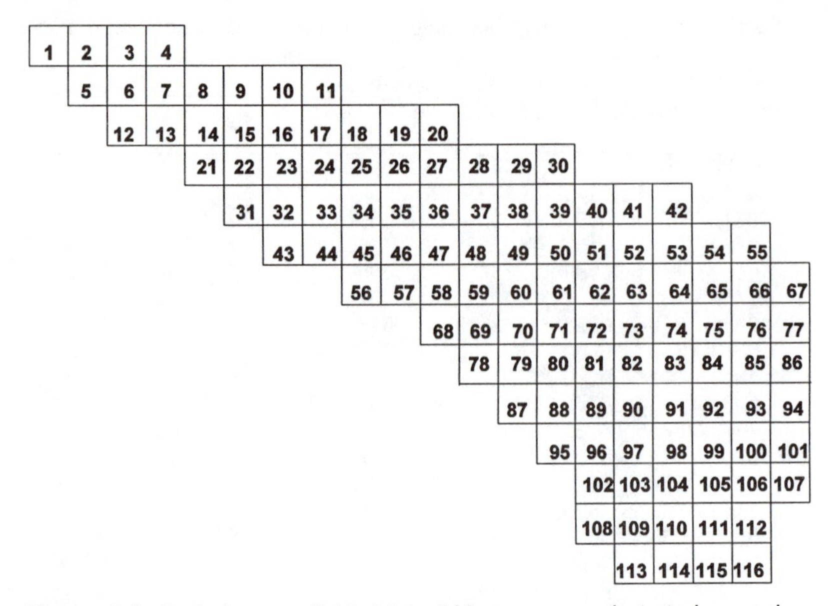

Figure 2.1 A study area divided into 116 square quadrats to be used as sample units.

2.3 Estimation of Population Means

Assume that a simple random sample of size n is selected without replacement from a population of N units, and that the variable of interest has values $y_1, y_2, \ldots, y_n$, for the sampled units. Then the sample mean is

$$\bar{y} = \sum_{i=1}^{n} y_i / n, \tag{2.1}$$

the sample variance is

$$s^2 = \{ \sum_{i=1}^{n} (y_i - \bar{y})^2 \}/(n - 1), \tag{2.2}$$

and the sample standard deviation is s, the square root of the variance. Equations (2.1) and (2.2) are the same as equations (A1) and (A2), respectively, in Appendix A except that the variable being considered is now labelled y instead of x. Another quantity that is sometimes of interest is the sample coefficient of variation

$$CV(y) = s/\bar{y}. \qquad (2.3)$$

These values that are calculated from samples are often referred to as sample statistics. The corresponding population values are the population mean μ, the population variance σ^2, the population standard deviation σ, and the population coefficient of variation σ/μ. These are often referred to as population parameters, and they are obtained by applying equations (2.1) to (2.3) to the full set of N units in the population. For example, μ is the mean of the observations on all of the N units.

The sample mean is an estimator of the population mean μ. The difference $\bar{y} - \mu$ is then the sampling error in the mean. This error will vary from sample to sample if the sampling process is repeated, and it can be shown theoretically that if this is done a large number of times then the error will average out to zero. For this reason the sample mean is said to be an unbiased estimator of the population mean.

It can also be shown theoretically that the distribution of $\bar{y}$ that is obtained by repeating the process of simple random sampling without replacement has the variance

$$Var(\bar{y}) = (\sigma^2/n)(1 - n/N). \qquad (2.4)$$

The factor {1 - n/N} is called the finite population correction because it makes an allowance for the size of the sample relative to the size of the population. The square root of $Var(\bar{y})$ is commonly called the standard error of the sample mean. It will be denoted here by $SE(\bar{y}) = \sqrt{Var(\bar{y})}$.

Because the population variance σ^2 will not usually be known it must usually be estimated by the sample variance s^2 for use in equation (2.4). The resulting estimate of the variance of the sample mean is then

$$\hat{Var}(\bar{y}) = \{s^2/n\}\{1 - n/N\}. \qquad (2.5)$$

The square root of this quantity is the estimated standard error of the mean

$$\hat{SE}(\bar{y}) = \sqrt{[\{s^2/n\}\{1 - n/N\}}. \qquad (2.6)$$

The 'caps' on $\hat{Var}(\bar{y})$ and $\hat{SE}(\bar{y})$ are used here to indicate estimated values, which is a common convention in statistics.

The terms 'standard error of the mean' and 'standard deviation' are often confused. What must be remembered is that the standard error of the mean is just the standard deviation of the mean rather than the standard deviation of individual observations. More generally, the term 'standard error' is used to describe the standard deviation of any sample statistic that is used to estimate a population parameter.

The accuracy of a sample mean for estimating the population mean is often represented by a $100(1-\alpha)\%$ confidence interval for the population mean of the form

$$\bar{y} \pm z_{\alpha/2} \hat{SE}(\bar{y}), \qquad (2.7)$$

where $z_{\alpha/2}$ refers to the value that is exceeded with probability $\alpha/2$ for the standard normal distribution, which can be determined using Table B1 if necessary. This is an approximate confidence interval for samples from any distribution, based on the result that sample means tend to be normally distributed even when the distribution being sampled is not. The interval is valid providing that the sample size is larger than about 25 and the distribution being sampled is not very extreme in the sense of having many tied values or a small proportion of very large or very small values.

Commonly used confidence intervals are

$$\bar{y} \pm 1.64\ \hat{SE}(\bar{y}) \text{ (90\% confidence)},$$
$$\bar{y} \pm 1.96\ \hat{SE}(\bar{y}) \text{ (95\% confidence), and}$$
$$\bar{y} \pm 2.58\ \hat{SE}(\bar{y}) \text{ (99\% confidence)}.$$

Often a 95% confidence interval is taken as $\bar{y} \pm 2\ \hat{SE}(\bar{y})$ on the grounds of simplicity, and because it makes some allowance for the fact that the standard error is only an estimated value.

The concept of a confidence interval is discussed in Section A5 of Appendix A. A 90% confidence interval is, for example, an interval within which the population mean will lie with probability 0.9. Put another way, if many such confidence intervals are calculated, then about 90% of these intervals will actually contain the population mean.

For samples that are smaller than 25 it is better to replace the confidence interval (2.7) with

$$\bar{y} \pm t_{\alpha/2,n-1} \hat{SE}(\bar{y}), \qquad (2.8)$$

where $t_{\alpha/2,n-1}$ is the value that is exceeded with probability $\alpha/2$ for the t-distribution with n-1 degrees of freedom. This is the interval that is justified in Section A5 of Appendix A samples from a normal distribution, except that the standard error used in that case was just $s/\sqrt{n}$ because a finite population correction was not involved. The use of the interval (2.8) requires the assumption that the variable being measured is approximately normally distributed in the population being sampled. It may not be satisfactory for samples from very non-symmetric distributions.

Example 2.1 Soil Percentage in the Corozal District of Belize

As part of a study of prehistoric land use in the Corozal District of Belize in Central America the area was divided into 151 plots of land with sides 2.5 by 2.5 km (Green, 1973). A simple random sample of 40 of these plots was selected without replacement, and provided the percentages of soils with constant lime enrichment that are shown in Table 2.2. This example considers the use of these data to estimate the average value of the measured variable (Y) for the entire area.

Table 2.2 Values for the percentage of soils with constant lime enrichment for 40 plots of land of size 2.5 by 2.5 km chosen by simple random sampling without replacement from 151 plots comprising the Corozal District of Belize in Central America

100	10	100	10	20	40	75	0	60	0
40	40	5	100	60	10	60	50	100	60
20	40	20	30	20	30	90	10	90	40
50	70	30	30	15	50	30	30	0	60

The mean percentage for the sampled plots is 42.38, and the standard deviation is 30.40. The estimated standard error of the mean is then found from equation (2.6) to be

$$\hat{SE}(\bar{y}) = \sqrt{[\{30.40^2/40\}\{1 - 40/151\}]} = 4.12.$$

Approximate 95% confidence limits for the population mean percentage are then found from equation (2.7) to be 42.38 ± 1.96x4.12, or 34.3 to 50.5.

In fact, Green (1973) provides the data for all 151 plots in his paper. The population mean percentage of soils with constant lime enrichment is therefore known to be 47.7%. This is well within the confidence limits, so the estimation procedure has been effective.

Note that the plot size used to define sample units in this example could have been different. A larger size would have led to a population with fewer sample units while a smaller size would have led to more sample units. The population mean, which is just the percentage of soils with constant lime enrichment in the entire study area, would be unchanged.

2.4 Estimation of Population Totals

In many situations there is more interest in the total of all values in a population, rather than the mean per sample unit. For example, the total area damaged by an oil spill is likely to be of more concern than the average area damaged on sample units. It turns out that the estimation of a population total is straightforward providing that the population size N is known, and an estimate of the population mean is available. It is obvious, for example, that if a population consists of 500 plots of land, with an estimated mean amount of oil spill damage of 15 square metres, then it is estimated that the total amount of damage for the whole population is 500 x 15 = 7500 square metres.

The general equation relating the population total T_y to the population mean μ for a variable Y is $T_y = N\mu$, where N is the population size. The obvious estimator of the total based on a sample mean $\bar{y}$ is therefore

$$t_y = N\bar{y}. \tag{2.9}$$

The sampling variance of this estimator is

$$\mathrm{Var}(t_y) = N^2\,\mathrm{Var}(\bar{y}), \tag{2.10}$$

and its standard error (i.e., standard deviation) is

$$\mathrm{SE}(t_y) = N\,\mathrm{SE}(\bar{y}). \tag{2.11}$$

Estimates of the variance and standard error are

$$\hat{\mathrm{Var}}(t_y) = N^2\,\hat{\mathrm{Var}}(\bar{y}), \tag{2.12}$$

and

$$\hat{SE}(t_y) = N\,\hat{SE}(\bar{y}). \tag{2.13}$$

In addition, an approximate 100(1-α)% confidence interval for the true population total can also be calculated in essentially the same manner as described in the previous section for finding a confidence interval for the population mean. Thus the limits are

$$t_y \pm z_{\alpha/2}\,\hat{SE}(t_y). \tag{2.14}$$

2.5 Estimation of Proportions

In discussing the estimation of population proportions it is important to distinguish between proportions measured on sample units and proportions of sample units. Proportions measured on sample units, such as the proportions of the units covered by a certain type of vegetation, can be treated like any other variables measured on the units. In particular, the theory for the estimation of the mean of a simple random sample that is covered in Section 2.3 applies for the estimation of the mean proportion. Indeed, Example 2.1 was of exactly this type except that the measurements on the sample units were percentages rather than proportions (i.e., proportions multiplied by 100). Proportions of sample units are different because the interest is in which units are of a particular type. An example of this situation is where the sample units are blocks of land and it is required to estimate the proportion of all the blocks that show evidence of damage from pollution. In this section only the estimation of proportions of sample units is considered.

Suppose that a simple random sample of size n, selected without replacement from a population of size N, contains r units with some characteristic of interest. Then the sample proportion is $\hat{p} = r/n$, and it can be shown that this has a sampling variance of

$$Var(\hat{p}) = \{p(1 - p)/n\}\{1 - n/N\}, \tag{2.15}$$

and a standard error of $SE(\hat{p}) = \sqrt{Var(\hat{p})}$. These results are the same as those obtained from assuming that r has a binomial distribution (see Appendix Section A2), but with a finite population correction.

Estimated values for the variance and standard error can be obtained by replacing the population proportion in equation (2.15) with the sample proportion $\hat{p}$. Thus the estimated variance is

$$V\hat{a}r(\hat{p}) = [\{\hat{p}(1 - \hat{p})/n\}\{1 - n/N\}], \tag{2.16}$$

and the estimated standard error is $\hat{SE}(\hat{p}) = \sqrt{\hat{Var}(\hat{p})}$. This creates little error in estimating the variance and standard error unless the sample size is quite small (say less than 20).

Using the estimated standard error, an approximate $100(1-\alpha)\%$ confidence interval for the true proportion is

$$\hat{p} \pm z_{\alpha/2}\, \hat{SE}(\hat{p}), \tag{2.17}$$

where, as before, $z_{\alpha/2}$ is the value from the standard normal distribution that is exceeded with probability $\alpha/2$.

The confidence limits produced by equation (2.17) are based on the assumption that the sample proportion is approximately normally distributed, which it will be if $np(1-p) \geq 5$ and the sample size is fairly small in comparison to the population size. If this is not the case, then alternative methods for calculating confidence limits should be used (Cochran, 1977, Section 3.6).

Example 2.2 PCB Concentrations in Surface Soil Samples

As an example of the estimation of a population proportion, consider some data provided by Gore and Patil (1994) on polychlorinated biphenyl (PCB) concentrations in parts per million (ppm) at the Armagh compressor station in West Wheatfield Township, along the gas pipeline of the Texas Eastern Pipeline Gas Company in Pennsylvania, USA. The cleanup criterion for PCB in this situation for a surface soil sample is an average PCB concentration of 5 ppm in soils between the surface and six inches in depth.

In order to study the PCB concentrations at the site, grids were set surrounding four potential sources of the chemical, with 25 feet separating the grid lines for the rows and columns. Samples were then taken at 358 of the points where the row and column grid lines intersected. Gore and Patil give the PCB concentrations at all of these points. However, here the estimation of the proportion of the N = 358 points at which the PCB concentration exceeds 5 ppm will be considered, based on a random sample of n = 100 of the points, selected without replacement.

The PCB values for the sample of 50 points are shown in Table 2.3. Of these, 31 exceed 5 ppm so that the estimate of the proportion of exceedances for all 358 points is $\hat{p} = 31/50 = 0.62$. The estimated variance associated with this proportion is then found from equation (2.16) to be

$$\hat{Var}(\hat{p}) = \{0.62 \times 0.38/50\}(1 - 50/358) = 0.0041.$$

Thus $\hat{SE}(\hat{p}) = 0.064$, and the approximate confidence interval for the proportion for all points, calculated from equation (2.17), is 0.495 to 0.745.

Table 2.3 PCB concentrations in parts per million at 50 sample points from the Armagh compressor station

5.1	49.0	36.0	34.0	5.4	38.0	1000.0	2.1	9.4	7.5
1.3	140.0	1.3	75.0	0.0	72.0	0.0	0.0	14.0	1.6
7.5	18.0	11.0	0.0	20.0	1.1	7.7	7.5	1.1	4.2
20.0	44.0	0.0	35.0	2.5	17.0	46.0	2.2	15.0	0.0
22.0	3.0	38.0	1880.0	7.4	26.0	2.9	5.0	33.0	2.8

2.6 Sampling and Non-Sampling Errors

Four sources of error may affect the estimation of population parameters from samples:

- Sampling errors are due to the variability between sample units and the random selection of units included in a sample.
- Measurement errors are due to the lack of uniformity in the manner in which a study is conducted, and inconsistencies in the measurement methods used.
- Missing data are due to the failure to measure some units in the sample.
- Errors of various types may be introduced in coding, tabulating, typing and editing data.

The first of these errors is allowed for in the usual equations for variances. Also, random measurement errors from a distribution with a mean of zero will just tend to inflate sample variances, and will therefore be accounted for along with the sampling errors. Therefore, the main concerns with sampling should be potential bias due to measurement errors that tend to be in one direction, missing data values that tend to be different from the known values, and errors introduced while processing data.

The last three types of error are sometimes called non-sampling errors. It is very important to ensure that these errors are minimal, and to appreciate that unless care is taken they may swamp the sampling errors that are reflected in variance calculations. This has been well recognized by environmental scientists in the last 15 years or so, with much attention given to the development of appropriate procedures for quality assurance and quality control (QA/QC). These matters are discussed by Keith (1991, 1996) and Liabastre *et al.* (1992), and are also a key element in the data quality objectives (DQO) process that is discussed in Section 2.15.

2.7 Stratified Random Sampling

A valid criticism of simple random sampling is that it leaves too much to chance, so that the number of sampled units in different parts of the population may not match the distribution in the population. One way to overcome this problem while still keeping the advantages of random sampling is to use stratified random sampling. This involves dividing the units in the population into non-overlapping strata, and selecting an independent simple random sample from each of these strata.

Often there is little to lose by using this more complicated type of sampling but there are some potential gains. First, if the individuals within strata are more similar than individuals in general, then the estimate of the overall population mean will have a smaller standard error than can be obtained with the same simple random sample size. Second, there may be value in having separate estimates of population parameters for the different strata. Third, stratification makes it possible to sample different parts of a population in different ways, which may make some cost savings possible.

However, stratification can also cause problems that are best avoided if possible. This was the case with two of the *Exxon Valdez* studies that were discussed in Example 1.1. Exxon's Shoreline Ecology Program and the Oil Spill Trustees' Coastal Habitat Injury Assessment were both upset to some extent by an initial misclassification of units to strata which meant that the final samples within the strata were not simple random samples. The outcome was that the results of these studies either require a rather complicated analysis or are susceptible to being discredited. The first problem that can occur is therefore that the stratification used may end up being inappropriate.

Another potential problem with using stratification is that after the data are collected using one form of stratification there is interest in analysing the results using a different stratification that was not foreseen in advance, or using an analysis that is different from the original one proposed. Because of the many different groups interested in environmental matters from different points of view this is always a possibility, and it led Overton and Stehman (1995) to argue strongly in favour of using simple sampling designs with limited or no stratification.

If stratification is to be employed, then generally this should be based on obvious considerations such as spatial location, areas within which the population is expected to be uniform, and the size of sampling units. For example, in sampling vegetation over a large area it is natural to take a map and partition the area into a few apparently homogeneous strata based on factors such as altitude and vegetation type. Usually the choice of how to stratify is just a question of common sense.

Assume that K strata have been chosen, with the ith of these having size N_i and the total population size being $\sum N_i = N$. Then if a random sample with size n_i is taken from the ith stratum the sample mean $\bar{y}_i$ will be an unbiased estimate of the true stratum mean μ_i, with estimated variance

$$\hat{\mathrm{Var}}(\bar{y}_i) = (s_i^2/n_i)(1 - n_i/N_i), \tag{2.18}$$

where s_i is the sample standard deviation within the stratum. These results follow by simply applying the results discussed earlier for simple random sampling to the ith stratum only.

In terms of the true strata means, the overall population mean is the weighted average

$$\mu = \sum_{i=1}^{K} N_i\mu_i/N, \tag{2.19}$$

and the corresponding sample estimate is

$$\bar{y}_s = \sum_{i=1}^{K} N_i\bar{y}_i/N, \tag{2.20}$$

with estimated variance

$$\hat{Var}(\bar{y}_s) = \sum_{i=1}^{K} (N_i/N)^2 \hat{Var}(\bar{y}_i)$$

$$= \sum_{i=1}^{K} (N_i/N)^2(s_i^2/n_i)(1 - n_i/N_i). \quad (2.21)$$

The estimated standard error of $\bar{y}_s$ is $\hat{SE}(\bar{y}_s)$, the square root of the estimated variance, and an approximate $100(1-\alpha)\%$ confidence interval for the population mean is given by

$$\bar{y}_s \pm z_{\alpha/2}\, \hat{SE}(\bar{y}_s), \quad (2.22)$$

where $z_{\alpha/2}$ is the value exceeded with probability $\alpha/2$ for the standard normal distribution.

If the population total is of interest, then this can be estimated by

$$t_s = N\bar{y}_s \quad (2.23)$$

with estimated standard error

$$\hat{SE}(t_s) = N\, \hat{SE}(\bar{y}_s). \quad (2.24)$$

Again, an approximate $100(1-\alpha)\%$ confidence interval takes the form

$$t_s \pm z_{\alpha/2}\, \hat{SE}(t_s). \quad (2.25)$$

Equations are available for estimating a population proportion from a stratified sample (Scheaffer *et al.*, 1990, Section 5.6). However, if an indicator variable Y is defined which takes the value one if a sample unit has the property of interest, and zero otherwise, then the mean of Y in the population is equal to the proportion of the sample units in the population that have the property. Therefore, the population proportion of units with the property can be estimated by applying equation (2.20) with the indicator variable, together with the equations for the variance and approximate confidence limits.

When a stratified sample of points in a spatial region is carried out it often will be the case that there are an unlimited number of sample points that can be taken from any of the strata, so that N_i and N are infinite. Equation (2.20) can then be modified to

$$\bar{y}_s = \sum_{i=1}^{K} w_i \bar{y}_i, \tag{2.26}$$

where w_i, the proportion of the total study area within the ith stratum, replaces N_i/N. Similarly, equation (2.21) changes to

$$\hat{Var}(\bar{y}_s) = \sum_{i=1}^{K} w_i^2 s_i^2/n_i. \tag{2.27}$$

Equations (2.22) to (2.25) remain unchanged.

Example 2.3 Bracken Density in Otago

As part of an ongoing study of the distribution of scrub weeds in New Zealand, data were obtained on the density of bracken on one hectare (100m by 100m) pixels along a transect 90km long and 3km wide running from Balclutha to Katiki Point on the South Island of New Zealand, as shown in Figure 2.2 (Gonzalez and Benwell, 1994). This example involves a comparison between estimating the density (the percentage of the land in the transect covered with bracken) using (i) a simple random sample of 400 pixels, and (ii) using a stratified random sample with five strata and the same total sample size.

There are altogether 27,000 pixels in the entire transect, most of which contain no bracken. The simple random sample of 400 pixels was found to contain 377 with no bracken, 14 with 5% bracken, 6 with 15% bracken, and 3 with 30% bracken. The sample mean is therefore $\bar{y}$ = 0.625%, the sample standard deviation is s = 3.261, and the estimated standard error of the mean is

$$\hat{SE}(\bar{y}_s) = (3.261/\sqrt{400})(1 - 400/27000) = 0.162.$$

The approximate 95% confidence limits for the true population mean density is therefore 0.625 ± 1.96 x 0.162, or 0.31 to 0.94%.

The strata for stratified sampling were five stretches of the transect, each about 18km long, and each containing 5400 pixels. The sample results and some of the calculations for this sample are shown in Table 2.4. The estimated population mean density from equation (2.19) is 0.613%, with an estimated variance of 0.0208 from equation (2.21). The estimated standard error is therefore $\sqrt{0.0208}$ = 0.144, and an approximate 95% confidence limits for the true population mean density is 0.613 ± 1.96 x 0.144, or 0.33 to 0.90%.

In a situation being considered there might be some interest in estimating the area in the study region covered by bracken. The total area is 27,000 hectares. Therefore the estimate from simple random sampling is 27,000 x 0.00625 = 168.8 hectares, with an estimated standard error of 27,000 x 0.00162 = 43.7 hectares, expressing the estimated percentage cover as a proportion. The approximate 95% confidence limits are 27,000 x 0.0031 = 83.7 to 27,000 x 0.0094 = 253.8 hectares. Similar calculations with the results of the stratified sample give an estimated coverage of 165.5 hectares, with a standard error of 38.9 hectares, and approximate 95% confidence limits of 89.1 to 243.0 hectares.

In this example the advantage of using stratified sampling instead of simple random sampling has not been great. The estimates of the mean bracken density are quite similar and the standard error from the stratified sample (0.144) is not much smaller than that for simple random sampling (0.162). Of course, if it had been known in advance that no bracken would be recorded in stratum 5, then the sample units in that stratum could have been allocated to the other strata, leading to some further reduction in the standard error. Methods for deciding on sample sizes for stratified and other sampling methods are discussed further in Section 2.13.

2.8 Post-Stratification

At times there may be value in analysing a simple random sample as if it were obtained by stratified random sampling. That is to say, a simple random sample is taken and the units are then placed into strata, possibly based on information obtained at the time of sampling. The sample is then analysed as if it were a stratified random sample in the first place, using the equations given in the previous section. This procedure is called post-stratification. It requires that the strata sizes N_i are known so that equations (2.20) and (2.21) can be used.

A simple random sample is expected to place sample units in different strata according to the size of those strata. Therefore, post-stratification should be quite similar to stratified sampling with proportional allocation, providing that the total sample size is reasonably large. It therefore has some considerable potential merit as a method that permits the method of stratification to be changed after a sample has been selected. This may be particularly valuable in situations where the data may be used for a variety of purposes, some of which are not known at the time of sampling.

Figure 2.2 A transect about 90km long and 3km wide along which bracken has been sampled in the South Island of New Zealand.

Table 2.4 The results of stratified random sampling for estimating the density of bracken along a transect in the South Island of New Zealand

	Stratum										
Case	1		2		3		4		5		
1	0	0	15	5	0	0	0	0	0	0	
2	0	0	0	0	0	0	0	0	0	0	
3	0	0	0	0	0	0	0	0	0	0	
4	0	0	0	0	0	0	0	0	0	0	
5	0	0	15	0	5	0	0	0	0	0	
6	0	0	0	0	0	0	0	0	0	0	
7	0	0	0	5	0	5	0	5	0	0	
8	0	0	0	0	0	0	0	0	0	0	
9	0	0	0	0	0	0	0	0	0	0	
10	0	0	0	0	0	0	0	0	0	0	
11	0	0	0	0	0	0	0	0	0	0	
12	0	0	0	0	0	0	0	0	0	0	
13	0	0	0	0	0	0	0	0	0	0	
14	0	0	5	0	0	0	0	30	0	0	
15	0	0	0	15	15	0	0	0	0	0	
16	0	0	5	0	0	30	0	0	0	0	
17	5	0	0	0	0	0	0	0	0	0	
18	0	0	15	0	0	0	0	0	0	0	
19	0	0	0	0	0	5	0	0	0	0	
20	0	0	0	0	0	0	0	0	0	0	
21	0	0	0	0	0	0	0	0	0	0	
22	0	0	0	0	0	0	0	0	0	0	
23	0	0	5	0	0	0	0	0	0	0	
24	0	0	0	0	0	0	0	0	0	0	
25	0	5	5	5	0	0	0	0	0	0	
26	0	0	0	0	0	0	0	0	0	0	
27	0	0	0	0	0	5	0	0	0	0	
28	0	0	0	0	0	0	0	0	0	0	
29	0	0	0	0	0	0	0	0	0	0	
30	0	0	5	5	0	0	0	0	0	0	
31	0	0	0	0	0	0	0	0	0	0	
32	0	0	15	5	0	0	0	0	0	0	
33	0	0	0	0	0	0	0	0	0	0	
34	0	0	0	0	0	0	0	0	0	0	
35	0	0	0	0	0	0	0	0	0	0	
36	0	0	0	0	0	0	0	0	0	0	
37	0	0	0	0	0	0	0	0	0	0	
38	0	0	0	0	0	0	0	0	0	0	
39	0	0	0	0	0	0	0	0	0	0	
40	0	5	5	0	0	0	0	0	0	0	
Mean	0.1875		1.625		0.8125		0.4375		0.000		
SD	0.956		3.879		3.852		3.393		0.000		Total
n	80		80		80		80		80		400
N	5400		5400		5400		5400		5400		27000

Contributions to the sum in equation (2.19) for the estimated mean

0.0375 0.3250 0.1625 0.0875 0.0000 0.6125

Contributions to the sum in equation (2.21) for the estimated variance

0.0005 0.0074 0.0073 0.0057 0.0000 0.0208

2.9 Systematic Sampling

Systematic sampling is often used as an alternative to simple random sampling or stratified random sampling for two reasons. First, the process of selecting sample units is simpler for systematic sampling. Second, under certain circumstances estimates can be expected to be more precise for systematic sampling because the population is covered more evenly.

The basic idea with systematic sampling is to take every kth item in a list, or to sample points that are regularly placed in space. As an example, consider the situation that is shown in Figure 2.3. The top part of the figure shows the positions of 12 randomly placed sample points in a rectangular study area. The middle part shows a stratified sample where the study region is divided into four equal sized strata, and three sample points are placed randomly within each. The lower part of the figure shows a systematic sample where the study area is divided into 12 equal sized quadrats each of which contains a point at the same randomly located position within the quadrat. Quite clearly, stratified sampling has produced better control than random sampling in terms of the way that the sample points cover the region, but not as much control as systematic sampling.

It is common to analyse a systematic sample as if it were a simple random sample. In particular, population means, totals and proportions are estimated using the equations in Sections 2.3 to 2.5, including the estimation of standard errors and the determination of confidence limits. The assumption is then made that because of the way that the systematic sample covers the population this will, if anything, result in standard errors that tend to be somewhat too large and confidence limits that tend to be somewhat too wide. That is to say, the assessment of the level of sampling errors is assumed to be conservative.

The only time that this procedure is liable to give a misleading impression about the true level of sampling errors is when the population being sampled has some cyclic variation in observations so that the regularly spaced observations that are selected tend to all be either higher or lower than the population mean. Therefore, if there is a suspicion that regularly spaced sample points may follow some pattern in the population values, then systematic sampling should be avoided. Simple random sampling and stratified random sampling are not affected by any patterns in the population, and it is therefore safer to use them when patterns may be present.

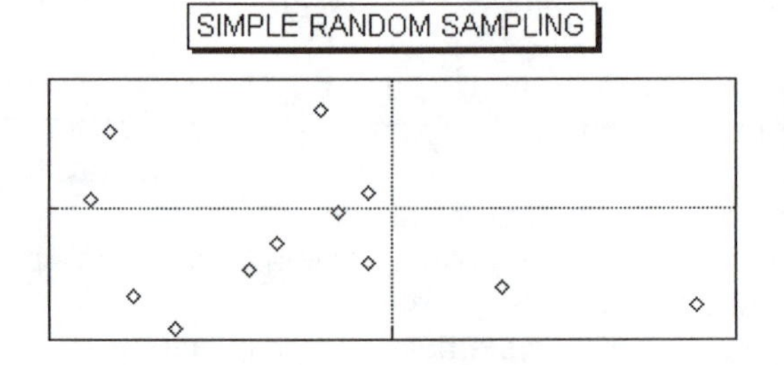

STRATIFIED RANDOM SAMPLING

SYSTEMATIC SAMPLING

Figure 2.3 Comparison of simple random sampling, stratified random sampling and systematic sampling for points in a rectangular study region.

The United States Environmental Protection Agency (1989a) manual on statistical methods for evaluating the attainment of site cleanup standards recommends two alternatives to treating a systematic sample as a simple random sample for the purpose of analysis. The first of these alternatives involves combining adjacent points into strata, as indicated in Figure 2.4. The population mean and standard error are then estimated using equations (2.26) and (2.27). The assumption being made is that the sample within each of the imposed strata is equivalent to a random sample. It is most important that the strata are defined without taking any notice of the values of observations because otherwise bias will be introduced into the variance calculation.

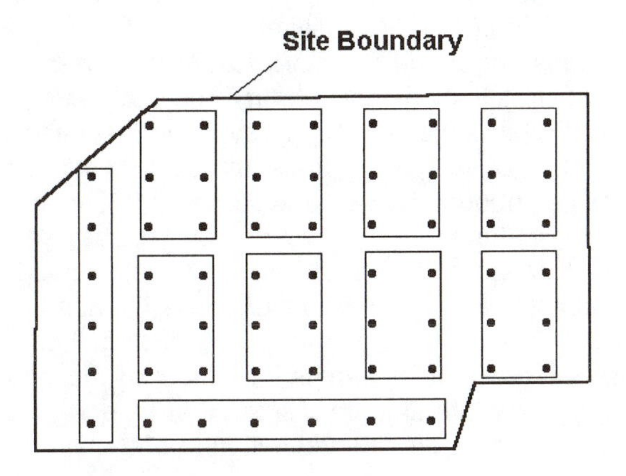

Figure 2.4 Grouping sample points from a systematic sample so that it can be analysed as a stratified sample. The sample points (•) are grouped here into 10 strata each containing six points.

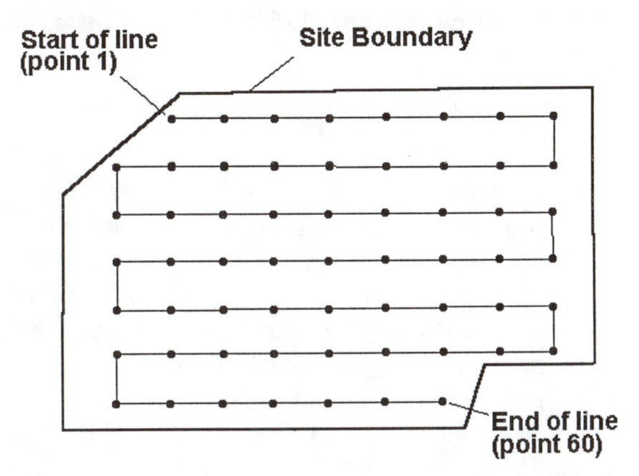

Figure 2.5 Defining a serpentine line connecting the points of a systematic sample so that the sampling variance can be estimated using squared differences between adjacent points on the line.

If the number of sample points or the area is not the same within each of the strata, then the estimated mean from equation (2.26) will differ from the simple mean of all of the observations. This is to be avoided because it will be an effect that is introduced by a more or less arbitrary system of stratification. The estimated variance of the mean from equation (2.27) will inevitably depend on the stratification used and under some circumstances it may be necessary to show that all reasonable stratifications give about the same result.

The second alternative to treating a systematic sample as a simple random sample involves joining the sample points with a serpentine line that joins neighbouring points and passes only once through each point, as shown in Figure 2.5. Assuming that this has been done, and that y_i is the ith observation along the line, it is assumed that y_{i-1} and y_i are both measures of the variable of interest in approximately the same location. The difference squared $(y_i - y_{i-1})^2$ is then an estimate of twice the variance of what can be thought of as the local sampling errors. With a systematic sample of size n there are n-1 such squared differences, leading to a combined estimate of the variance of local sampling errors of

$$s_L^2 = \tfrac{1}{2} \sum_{i=2}^{n} (y_i - y_{i-1})^2/(n-1). \qquad (2.28)$$

On this basis the estimate of the standard error of the mean of the systematic sample is

$$\hat{SE}(\bar{y}) = s_L/\sqrt{n}. \qquad (2.29)$$

No finite sampling correction is applied when estimating the standard error on the presumption that the number of potential sampling points in the study area is very large. Once the standard error is estimated using equation (2.29) approximate confidence limits can be determined using equation (2.7), and the population total can be estimated using the methods described in Section 2.4. This approach for assessing sampling errors was found to be as good or better than seven alternatives from a study that was carried out by Wolter (1984).

Example 2.4 Total PCBs in Liverpool Bay Sediments

Camacho-Ibar and McEvoy (1996) describe a study that was aimed at determining the concentrations of 55 polychlorinated biphenyl (PCB) congeners in sediment samples from Liverpool Bay in the

United Kingdom. For this purpose, the total PCB was defined as the summation of the concentrations of all the identifiable individual congeners that were present at detectable levels in each of 66 grab samples taken between 14 and 16 September 1988. The values for this variable and the approximate position of each sample are shown in Figure 2.6. Although the sample locations were not systematically placed over the bay, they are much more regularly spaced than can be expected to occur with random sampling.

The mean and standard deviation of the 66 observations are $\bar{y}$ = 3937.7 and s = 6646.5, in units of pg g^{-1} (picograms per gram, i.e., parts per 10^{12}). Therefore, if the sample is treated as being equivalent to a simple random sample, then the estimated standard error is $\hat{SE}(\bar{y})$ = 6646.5/√66 = 818.1, and the approximate 95% confidence limits for the mean over the sampled region are 3937.7 ± 1.96 x 818.1, or 2334.1 to 5541.2.

The second method for assessing the accuracy of the mean of a systematic sample as described above entails dividing the samples into strata. This division was done arbitrarily using 11 strata of six observations each, as shown in Figure 2.7, and the calculations for the resulting stratified sample are shown in Table 2.5. The estimated mean level for total PCBs in the area is still 3937.7 pg g^{-1}. However, the standard error calculated from the stratification is 674.2, which is lower than the value of 818.1 found by treating the data as coming from a simple random sample. The approximate 95% confidence limits from stratification are 3937.7 ± 1.96 x 674.2, or 2616.2 to 5259.1.

Finally, the standard error can be estimated using equations (2.28) and (2.29), with the sample points in the order shown in Figure 2.7 but with the closest points connected between the sets of six observations that formed the strata before. This produces an estimated standard deviation of s_L = 5704.8 for small-scale sampling errors, and an estimated standard error for the mean in the study area of $\hat{SE}(\bar{y})$ = 5704.8/√66 = 702.2. By this method the approximate 95% confidence limits for the area mean are 3937.7 ± 1.96 x 702.2, or 2561.3 to 5314.0 pg g^{-1}. This is quite close to what was obtained using the stratification method.

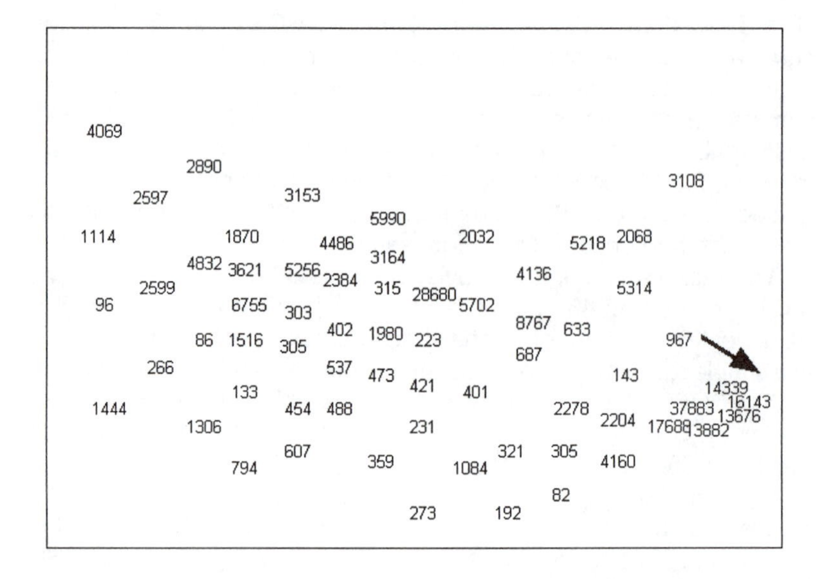

Figure 2.6 Concentration of total PCBs (pg g^{-1}) in samples of sediment taken from Liverpool Bay. Observations are shown at their approximate position in the study area. The arrow points to the entrance to the River Mersey.

2.10 Other Design Strategies

So far in this chapter the sample designs that have been considered are simple random sampling, stratified random sampling, and systematic sampling. There are also a number of other design strategies that are sometimes used. Here some of the designs that may be useful in environmental studies are just briefly mentioned. For further details see Scheaffer *et al.* (1990), Thompson (1992), or some other specialized text.

With cluster sampling, groups of sample units that are close in some sense are randomly sampled together, and then all measured. The idea is that this will reduce the cost of sampling each unit so that more units can be measured than would be possible if they were all sampled individually. This advantage is offset to some extent by the tendency of sample units that are close together to have similar measurements. Therefore, in general, a cluster sample of n units will give estimates that are less precise than a simple random sample of n units. Nevertheless, cluster sampling may give better value for money than the sampling of individual units.

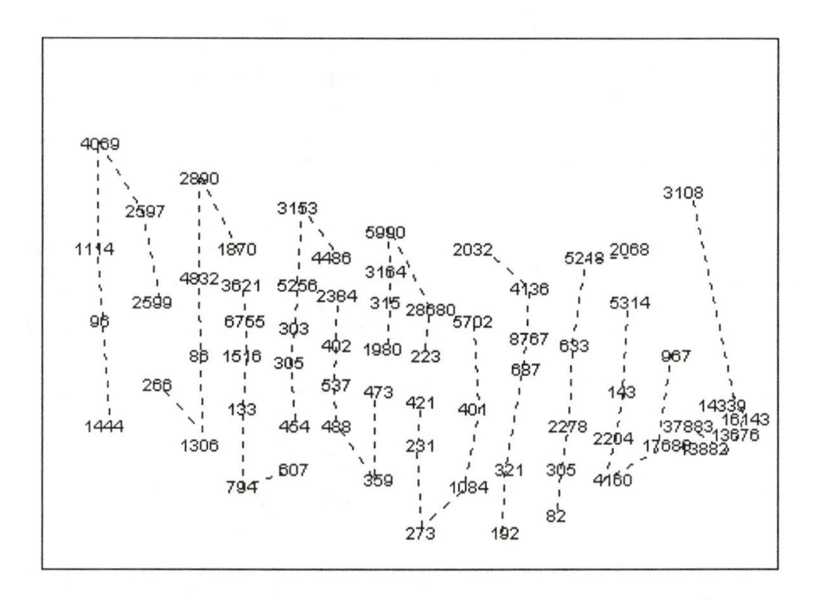

Figure 2.7 Partitioning of samples in Liverpool Bay into 11 strata consisting of points that are connected by broken lines.

With multi-stage sampling, the sample units are regarded as falling within a hierarchic structure. Random sampling is then conducted at the various levels within this structure. For example, suppose that there is interest in estimating the mean of some water quality variable in the lakes in a very large area such as a whole country. The country might then be divided into primary sampling units consisting of states or provinces, each primary unit might then consist of a number of counties, and each county might contain a certain number of lakes. A three-stage sample of lakes could then be obtained by first randomly selecting several primary sampling units, next randomly selecting one or more counties (second-stage units) within each sampled primary unit, and finally randomly selecting one or more lakes (third-stage units) from each sampled county. This type of sampling plan may be useful when a hierarchic structure already exists, or when it is simply convenient to sample at two or more levels.

Table 2.5 Treating the data on total PCBs as coming from a stratified sample. The 66 sample points have been grouped into 11 strata with six points in each (Figure 2.7). The last column in the table shows the contributions from each of the strata to the summation on the right-hand side of equation (2.27) for the stratified sample variance. For this equation all strata are treated as being of equal size so that $w_i = 1/11$ for all i. The standard error of the estimated mean (SE) is the square root of the sum of the last column

Stratum	Total PCB (pg g^{-1})						Mean	SD	Contribution to variance
1	1444	96	1114	4069	2597	2599	1986.5	1393.9	2676.1
2	266	1306	86	4832	2890	1870	1875.0	1782.5	4376.7
3	3621	6755	1516	133	794	607	2237.7	2530.4	8819.2
4	454	305	303	5256	3153	4486	2326.2	2263.4	7056.7
5	2384	402	537	488	359	473	773.8	791.4	862.6
6	1980	315	3164	5990	28680	223	6725.3	10964.7	165598.4
7	421	231	273	1084	401	5702	1352.0	2153.4	6387.4
8	2032	4136	8767	687	321	192	2689.2	3326.2	15239.1
9	82	305	2278	633	5218	2068	1764.0	1924.8	5102.9
10	5314	143	2204	4160	17688	967	5079.3	6471.9	57692.8
11	3108	14339	16143	37883	13676	13882	16505.2	11456.0	180772.6
							3937.7		454584.5
									SE = 674.2

The technique of composite sampling is valuable in situations where the cost of selecting and acquiring sample units is much less than the cost of analysing them. What this involves is mixing several samples from approximately the same location and then analysing the composite samples. For example, sets of four samples might be mixed so that the number of analyses is only one quarter of the number of samples. This should have little effect on the estimated mean providing that samples are mixed sufficiently so that the observation from a composite sample is close to the mean for the samples that it contains. However, there is a loss of information about extreme values for individual sample units because of dilution effects. If there is a need to identify individual samples with extreme values, then methods are available to achieve this without the need to analyse every sample. These and other aspects of composite sampling are discussed by Gore and Patil (1994) and Patil (1995), while Gilbert (1987) considers the estimation of the mean when composite samples are used with more complicated sampling designs.

Ranked set sampling is another method that can be used to reduce the cost of analysis in surveys. The technique was originally developed for the estimation of the biomass of vegetation (McIntyre,

1952), but the potential uses are much wider. It relies on the existence of an inexpensive method of assessing the relative magnitude of a small set of observations to supplement expensive accurate measurements.

As an example, suppose that 90 uniformly spaced sample units are arranged in a rectangular grid over an intertidal study area, and that it is necessary to estimate the average barnacle density. A visual assessment is made of the density on the first three units, which are then on that basis ordered from the one with the lowest density to the one with the highest density. The density is then determined accurately for the highest ranked unit. The next three units are then visually ranked in the same way, and the density is then determined accurately for the unit with the middle of the three ranks. Next, sample units 7, 8, and 9 are ranked, and the density determined accurately for the unit with the lowest rank. The process of visually ranking sets of three units and measuring first the highest ranking unit, then the middle ranking unit, and finally the lowest ranking unit is then repeated using units 10 to 18, units 19 to 27, and so on. After the completion of this procedure on all 90 units a ranked set sample of size 30 is available based on the accurate estimation of density. This sample is not as good as would have been obtained by measuring all 90 units accurately, but it should have considerably better precision than a standard sample of size 30.

Let m be the size of the ranked set (3 in the above example), so that one cycle of ranking and measurement uses m^2 units, accurately measuring first the highest of m units then the second highest of m units, and so on. Also, let k be the number of times that the cycle is repeated (10 in the above example), so that $n = km$ values are accurately measured in total. Then it can be shown that if there are no errors in the ranking of the sets of size m and the distribution of the variable being measured is unimodal, the ratio of the variance of the mean from ranked set sampling to the variance of the mean for a simple random sample of size n is slightly more than $2/(m + 1)$. Thus in the example with $m = 3$ and a ranked set sample size of 30 the variance of the sample mean should be slightly more than $2/(3+1) = 1/2$ of the variance of a simple random sample of size 30.

In practice the user of ranked set sampling has two alternatives in terms of assessing the error in estimating a population mean using the mean of the ranked set sample. If there are few if any errors in ranking, and if the distribution of data values in the population being studied can be assumed to be unimodal, then the standard error of the ranked set sample mean can be estimated approximately by

$$\hat{SE}(\bar{x}) = \sqrt{\{[2/(m + 1)][s^2/n]\}}. \quad (2.30)$$

Alternatively, the sample can conservatively be treated as being equivalent to a simple random sample of size n, in which case the estimated standard error

$$\hat{SE}(\bar{x}) = s/\sqrt{n} \quad (2.31)$$

will almost certainly be too large.

For a further discussion and more details about ranked set sampling see the review by Patil *et al.* (1994), and the special issue of the journal *Environmental and Ecological Statistics* on this topic (Ross and Stokes, 1999).

2.11 Ratio Estimation

Occasions arise where the estimation of the population mean or total for a variable X is assisted by information on a subsidiary variable U. What is required is that the items in a random sample of size n have values x_1 to x_n for X, and corresponding values u_1 to u_n for U. In addition, μ_u, the population mean for U, and $T_u = N\mu_u$, the population total for U, must be known values. An example of such a situation would be where the level of a chemical was measured some years ago for all of the sample units in a population and it is required to estimate the current mean level from a simple random sample of the units. Then the level of the chemical from the earlier survey is the variable U and the current level is the variable X.

With ratio estimation it is assumed that X and U are approximately proportional, so that $X \approx RU$, where R is some constant. The value of the ratio R of X to U can then be estimated from the sample data by

$$r = \bar{x}/\bar{u}, \quad (2.32)$$

and hence the population mean of X can be estimated by multiplying r by the population mean of U, to get

$$\bar{x}_{ratio} = r\mu_u, \quad (2.33)$$

which is the ratio estimator of the population mean. Multiplying both sides of this equation by the population size N, the ratio estimate of the population total for X is found to be

$$t_X = rT_u. \quad (2.34)$$

If the ratio of X to U should be relatively constant for the units in a population, then the ratio estimator of the population mean for X can be expected to have lower standard errors than the sample mean $\bar{x}$. Similarly, the ratio estimator of the population total for X should have a lower standard error than $N\bar{x}$. This is because the ratio estimators allow for the fact that the observed sample may, by chance alone, consist of items with rather low or high values for X. Even if the random sample does not reflect the population very well the estimate of r may still be reasonable, which is all that is required for a good estimate of the population total.

The variance of $\bar{x}_{ratio}$ can be estimated by

$$\hat{Var}(\bar{x}_{ratio}) \approx \sum_{i=1}^{n} (x_i - ru_i)^2/\{(n - 1)n\}(1 - n/N). \tag{2.35}$$

An approximate 100(1-α)% confidence interval for the population mean of X is then given by

$$\bar{x}_{ratio} \pm z_{\alpha/2}\, \hat{SE}(\bar{x}_{ratio}), \tag{2.36}$$

where $\hat{SE}(\bar{x}_{ratio}) = \sqrt{\hat{Var}(\bar{x}_{ratio})}$.

Because the ratio estimator of the population total of X is $t_x = N\bar{x}_{ratio}$, it also follows that

$$\hat{SE}(t_X) \approx N^2\hat{SE}(\bar{x}_{ratio}), \tag{2.37}$$

and an approximate 100(1-α)% confidence interval for the true total is given by

$$t_X \pm z_{\alpha/2}\, \hat{SE}(t_x). \tag{2.38}$$

The equations for variances, standard errors and confidence intervals should give reasonable results providing that the sample size n is large, (which in practice means 30 or more), and the coefficients of variation of $\bar{x}$ and $\bar{u}$ (the standard errors of these sample means divided by their population means) to be less than 0.1 (Cochran, 1977, p. 153).

Ratio estimation assumes that the ratio of the variable of interest X to the subsidiary variable U is approximately constant for the items in the population. A less restrictive assumption is that X and U are related by an equation of the form $X \approx \alpha + \beta U$, where α and β are constants. This then allows regression estimation to be used. See

Manly (1992, Chapter 2) for more information about this generalization of ratio estimation.

Example 2.5 pH Levels in Norwegian Lakes

Example 1.2 was concerned with a Norwegian study that was started in 1972 to investigate the effects of acid precipitation. As part of this study, the pH levels of 68 lakes were measured in 1976, and for 32 of the lakes the pH level was measured again in 1977. The results are shown in Table 2.6 for the 32 lakes that were measured in both years. The present example is concerned with the estimation of the mean pH level for the population of 68 lakes in 1977 using the ratio and regression methods. Figure 2.8 shows a plot of the 1977 pH values against the 1976 pH values for the 32 lakes. Because a line through the data passes nearly through the origin there is approximately a ratio relationship as well so that ratio estimation is justified.

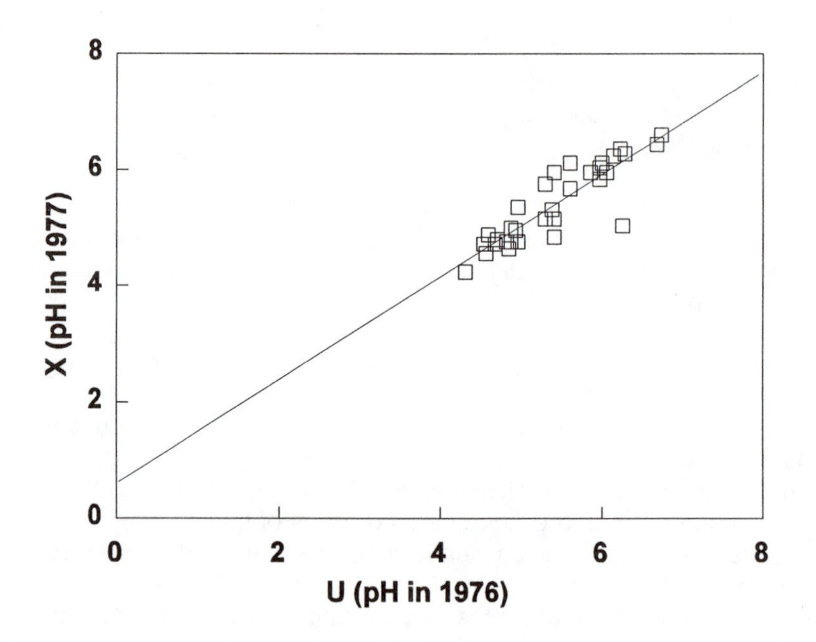

Figure 2.8 The relationship between pH values in 1976 and 1977 for 32 Norwegian lakes that were sampled in both years.

For ratio estimation, the pH level in a lake in 1976 is the auxiliary variable (U), and the pH level in 1977 is the main variable of interest (X). The mean of U is known to be 5.715 in 1976. For the 32 lakes sampled in 1977 the mean of U is $\bar{u} = 5.416$ and the mean of X is $\bar{x} = 5.400$. Therefore the estimated ratio of X to U is

$$r = 5.400/5.416 = 0.997.$$

The ratio estimate of the mean pH for all 68 lakes in 1977 is therefore given by equation (2.33) to be

$$\bar{x}_{ratio} = 0.997 \text{x} 5.715 = 5.70.$$

The column headed X - rU in Table 2.6 gives the values required for the summation on the right-hand side of equation (2.35). The sum of this column is zero and the value given as the standard deviation at the foot of the column (0.324) is the square root of $\sum(x_i - rU_i)^2/(n - 1)$. The estimated variance of the ratio estimator is therefore

$$\hat{Var}(\bar{x}_{ratio}) = (0.324^2/32)(1 - 32/68) = 0.00173,$$

and the estimated standard error is $\sqrt{0.00173} = 0.042$. An approximate 95% confidence interval for the mean pH in all lakes in 1977 is therefore 5.70 ± 1.96 x 0.042, or 5.62 to 5.78.

It is interesting to see that a rather different result is obtained if the mean pH value from the sample of 32 lakes is used to estimate the mean for all 68 lakes in 1977. The sample mean is 5.42, with an estimated standard error of 0.086. An approximate 95% confidence interval for the population mean is therefore 5.42 ± 1.96 x 0.086, or 5.25 to 5.59. Two remarks can be made here. First, the estimate of the population mean is now much lower than it was with ratio and regression estimation. Second, the standard error is much larger than it was with ratio and regression estimation.

The lower estimate for the population mean is likely to be the result of non-random sampling of the 68 lakes in 1977. The method for the selection of the lakes is not clear from Mohn and Volden's (1985) description, but appears to have been based on convenience to some extent. This approach is always, of course, liable to introduce a bias. It can be noted, however, that ratio estimation will not be biased unduly by the non-random selection of sample units providing that the units that are selected give good estimates of the ratio of X to U. In the situation being considered this suggests that ratio estimation may be more or less unbiased while estimation using the sample mean pH for 1977 is probably not.

Table 2.6 Values for pH measured on 32 Norwegian lakes in 1976 and 1977. The mean pH for the population of all 68 lakes measured in 1976 was 5.416. The lake identifier (Id) is as used by Mohn and Volden (1985)

Lake	Id	pH 1976 U	1977 X	X-rU
1	4	4.32	4.23	-0.077
2	5	4.97	4.74	-0.215
3	6	4.58	4.55	-0.016
4	8	4.72	4.81	0.104
5	9	4.53	4.70	0.184
6	10	4.96	5.35	0.405
7	11	5.31	5.14	-0.154
8	12	5.42	5.15	-0.254
9	17	4.87	4.76	-0.095
10	18	5.87	5.95	0.098
11	19	6.27	6.28	0.029
12	20	6.67	6.44	-0.210
13	24	5.38	5.32	-0.044
14	26	5.41	5.94	0.546
15	30	5.60	6.10	0.517
16	32	4.93	4.94	0.025
17	36	5.60	5.69	0.107
18	38	6.72	6.59	-0.110
19	40	5.97	6.02	0.068
20	41	4.68	4.72	0.054
21	43	6.23	6.34	0.129
22	47	6.15	6.23	0.098
23	49	4.82	4.77	-0.036
24	50	5.42	4.82	-0.584
25	58	5.31	5.77	0.476
26	59	6.26	5.03	-1.211
27	65	5.99	6.10	0.128
28	83	4.88	4.99	0.125
29	85	4.60	4.88	0.294
30	86	4.85	4.65	-0.185
31	88	5.97	5.82	-0.132
32	94	6.05	5.97	-0.062
	Mean	5.416	5.4	0
	SD	0.668	0.672	0.324

2.12 Double Sampling

In the previous section it was assumed that the value of the auxiliary variable U, or at least its population mean value, is known. Sometimes this is not the case and instead the following procedure is used. First, a large sample is taken and the value of U is measured on all of the sample units. Next, a small random subsample of the larger sample is selected, and the values of X are measured on this. The larger sample then provides an accurate estimate of the population mean of U which is used for ratio or regression estimation in place of the exact population mean of U.

This procedure will be useful if it is much easier or much less expensive to measure U than it is to measure X, providing that there is a good ratio or linear relationship between the two variables. A variation of this sample design can also be used with post-stratification of the larger sample.

The analysis of data from these types of double sampling designs is discussed by Scheaffer *et al.* (1990, Section 5.11) and Thompson (1992, Chapter 14). More details of the theory are provided by Cochran (1977, Chapter 12).

2.13 Choosing Sample Sizes

One of the most important questions for the design of any sampling programme is the total sample size that is required and, where it is relevant, how this total sample size should be allocated to different parts of the population. There are a number of specific aids available in the form of equations and tables that can assist in this respect. However, before considering these it is appropriate to mention a few general points.

First, it is worth noting that as a general rule the sample size for a study should be large enough so that important parameters are estimated with sufficient precision to be useful, but it should not be unnecessarily large. This is because on the one hand small samples with unacceptable levels of error are hardly worth doing at all while, on the other hand, very large samples giving more precision than is needed are a waste of time and resources. In fact, the reality is that the main danger in this respect is that samples will be too small. A number of authors have documented this in recent years in different areas of application. For example, Peterman (1990) describes several situations where important decisions concerning fisheries management have been based on the results of samples that were not adequate.

The reason why sample sizes tend to be too small if they are not considered properly in advance is that it is a common experience with environmental sampling that "what is desirable is not affordable, and what is affordable is not adequate" (Gore and Patil, 1994). There is no simple answer to this problem but researchers should at least know in advance if they are unlikely to achieve their desired levels of precision with the resources available, and possibly seek the extra resources that are needed. Also, it suggests that a reasonable strategy for determining sample sizes involves deciding what is the maximum size that is possible within the bounds of the resources available. The accuracy that can be expected from this size can then be assessed. If this accuracy is acceptable, but not as good as the researcher would like, then this maximum study size can be used on the grounds that it is the best that can be done. On the other hand, if a study of the maximum size gives an unnecessary level of accuracy, then the possibility of a smaller study can be investigated.

Another general approach to sample size determination that can usually be used fairly easily is trial and error. For example a spreadsheet can be set up to carry out the analysis that is intended for a study and the results of using different sample sizes can be explored using simulated data drawn from the type of distribution or distributions that are thought likely to occur in practice. A variation on this approach involves generation of data by bootstrap resampling of data from earlier studies. This involves generating test data by randomly sampling with replacement from the earlier data, as discussed by Manly (1992, p. 329). It has the advantage of not requiring arbitrary decisions about the distribution that the data will follow in the proposed new study.

Equations are available for determining sample sizes in some specific situations. Some results that are useful for large populations are as follows. For details of their derivation, and results for small populations, see Manly (1992, Section 11.4). In all cases δ represents a level of error that is considered to be acceptable by those carrying out a study. That is to say, δ is to be chosen by the investigators based on their objectives.

(a) To estimate a population mean from a simple random sample with a 95% confidence interval of $\bar{x} \pm \delta$, the sample size should be approximately

$$n = 4\sigma^2/\delta^2, \tag{2.39}$$

where σ is the population standard deviation. To use this equation, an estimate or guess of σ must be available.

(b) To obtain a 95% confidence limit for a population proportion of the form $p \pm \delta$, where p is the proportion in a simple random sample, requires that the sample size should be approximately

$$n = 4\pi(1-\pi)/\delta^2, \tag{2.40}$$

where π is the true population proportion. This has the upper limit of $n = 1/\delta^2$ when $\pi = \frac{1}{2}$, which gives a 'safe' sample size, whatever is the value of π.

(c) Suppose two random samples of size n are taken from different distributions, and give sample means of $\bar{x}_1$ and $\bar{x}_2$. Then to obtain an approximate 95% confidence interval for the difference between the two population means of the form $\bar{x}_1 - \bar{x}_2 \pm \delta$ requires that the sample sizes should be approximately

$$n = 8\sigma^2/\delta^2. \tag{2.41}$$

(d) Suppose that the difference between two sample proportions $\hat{p}_1$ and $\hat{p}_2$, with sample sizes of n, is to be used to estimate the difference between the corresponding population proportions p_1 and p_2. To obtain an approximate 95% confidence interval for the difference between the population proportions of the form $\hat{p}_1 - \hat{p}_2 \pm \delta$ requires that n is approximately

$$n \approx 8\pi'(1 - \pi')/\delta^2, \tag{2.42}$$

where π' is the average of the two population proportions. The largest possible value of n occurs with this equation when $\pi' = \frac{1}{2}$, in which case $n \approx 2/\delta^2$. This is therefore a 'safe' sample size for any population proportions.

The sample size equations just provided are based on the assumption that sample statistics are approximately normally distributed, and that sample sizes are large enough for the standard errors estimated from samples to be reasonably close to the true standard errors. In essence, this means that the sample sizes produced by the equations must be treated with some reservations unless there are at least 20 and the distribution or distributions being sampled are not grossly non-normal. Generally, the larger the sample size, the less important is the normality of the distribution being sampled.

For stratified random sampling it is necessary to decide on an overall sample size, and also how this should be allocated to the different strata. These matters are considered by Manly (1992, Section 2.7). In brief, it can be said the most efficient allocation to strata is one where n_i, the sample size in the ith stratum, is proportional to $N_i\sigma_i/\sqrt{c_i}$, where N_i is the size, σ_i is the standard deviation, and c_i is the cost of sampling one unit for this stratum. Therefore, when there is no reason to believe that the standard deviations vary greatly and sampling costs are about the same in all strata it is sensible to use proportional allocation, with n_i proportional to N_i.

Sample design and analysis can be a complex business, dependent very much on the particular circumstances (Rowan *et al.*, 1995; Lemeshow *et al.*, 1990; Borgman et al., 1996). There are now a number of computer packages for assisting with this task, such as PASS (Power Analysis and Sample Size), that is produced by NCSS Statistical Software (1995).

Sample size determination is an important component in the Data Quality Objectives (DQO) process that is described in Section 2.15.

2.14 Unequal Probability Sampling

The sampling theory discussed so far in this chapter is based on the assumption of random sampling of the population of interest. In other words, the population is sampled in such a way that each unit in the population has the same chance of being selected. It is true that this is modified with the more complex schemes involving, for example, stratified or cluster sampling, but even in these cases random sampling is used to choose units from those available.

However, situations do arise where the nature of the sampling mechanism makes random sampling impossible because the availability of sample units is not under the control of the investigator, so that there is unequal probability sampling. In particular, cases occur where the probability of a unit being sampled is a function of the characteristics of that unit. For example, large units might be more conspicuous than small ones, so that the probability of a unit being selected depends on its size. If the probability of selection is proportional to the size of units, then this special case is called size-biased sampling.

It is possible to estimate population parameters allowing for unequal probability sampling. Thus, suppose that the population being sampled contains N units, with values y_1, y_2, ... y_N for a variable Y, and that sampling is carried out so that the probability of including y_i in the sample is p_i. Assume that estimation of the population size (N), the population mean (μ_y), and the population total (T_y) is of interest, and that the sampling process yields n observed units. Then the population size can be estimated by

$$\hat{N} = \sum_{i=1}^{n} (1 / p_i), \qquad (2.43)$$

the total of Y can be estimated by

$$t_y = \sum_{i=1}^{n} (y_i / p_i), \qquad (2.44)$$

and the mean of Y can be estimated by

$$\hat{\mu}_y = t_y / \hat{N} = \sum_{i=1}^{n} (y_i / p_i) / \sum_{i=1}^{n} (1 / p_i). \qquad (2.45)$$

The estimators (2.43) and (2.44) are called Horvitz-Thomson estimators after those who developed them in the first place (Horvitz and Thompson, 1952). They provide unbiased estimates of the population parameters because of the weight given to different observations. For example, suppose that there are a number of population units with $p_i = 0.1$. Then it is expected that only one in ten of these units will appear in the sample of observed units. Consequently, the observation for any of these units that are observed should be weighted by $1/p_i = 10$ to account for those units that are missed from the sample.

Variance equations for all three estimators are provided by McDonald and Manly (1989), who suggest that replications of the sampling procedure will be a more reliable way of determining variances. Alternatively, bootstrapping may be effective. The book by Thomson (1992) gives a comprehensive guide to the many situations that occur where unequal probability sampling is involved.

2.15 The Data Quality Objectives Process

The Data Quality Objectives (DQO) process was developed by the United States Environmental Protection Agency (US EPA) to ensure that when a data collection process has been completed it will have:

- provided sufficient data to make required decisions within a reasonable certainty
- collected only the minimum amount of necessary data

The idea is to have the least expensive data collection scheme, but not at the price of providing answers that have too much uncertainty (United States Office of Environmental Management, 1997).

At the heart of the use of the process is the assumption that there will always be two problems with environmental decision making: (1) the resources available to address the question being considered are not infinite, and (2) there will never be a 100% guarantee that the right decision has been reached. Generally, more resources can be expected to reduce uncertainty. The DQO process therefore attempts to get the right balance between resource use and uncertainty, and provides a complete and defensible justification for the data collection methods used, covering:

- the questions that are important
- whether the data will answer the questions
- what quality of data is needed
- how much data are needed
- how the data will actually be used in decision making

This is all done before the data are collected, and preferably agreed to by all the stakeholders involved.

There are seven steps to the DQO process:

1. State the problem: describe the problem, review prior work, understand the important factors.

2. Identify the decision: find what questions need to be answered and the actions that might be taken, depending on the answers.

3. Identify inputs to the decision: determine the data needed to answer the important questions.

4. Define the study boundaries: specify the time periods and spatial areas to which decisions will apply, determine when and where to gather data.

5. Develop a decision rule: define the parameter of interest, specify action limits, integrate the previous DQO outputs into a single

statement that describes the logical basis for choosing among alternative possible actions.

6. Specify limits on decision errors: specify tolerable decision error probabilities (probabilities of making the wrong decisions) based on the consequences of incorrect decisions.

7. Optimize the design for obtaining data: generate alternative sampling designs, choose the one that meets all the DQOs with the minimum use of resources.

The output from each step influences the choices made later but it is important to realize that the process is iterative and the carrying out of one step may make it necessary to reconsider one or more earlier steps. Steps 1-6 should produce the Data Quality Objectives that are needed to develop the sampling design at step 7.

When used by the US EPA, a DQO planning team usually consists of technical experts, senior managers, a statistical expert and a quality assurance/quality control (QA/QC) advisor. The final product is a data collection design that meets the qualitative and quantitative needs of the study, and much of the information generated during the process is used for the development of Quality Assurance Project Plans (QAPPs) and the implementation of the Data Quality Assessment (DQA) Process. These are all part of the US EPA's system for maintaining quality in their operations.

More information about the DQO process with reference documents can be obtained from the world-wide web (United States Office of Environmental Management, 1997). A good start is to read the EPA's guidance document (United States Environmental Protection Agency, 1994).

2.16 Chapter Summary

- A crucial early task in any sampling study is to define the population of interest and the sample units that make up this population. This may or may not be straightforward.

- Simple random sampling involves choosing sample units in such a way that each unit is equally likely to be selected. It can be with or without replacement.

- Equations for estimating a population mean, total and proportion are provided for simple random sampling.

- Sampling errors are those due to the random selection of sample units for measurement. Non-sampling errors are those from all other sources. The control of both sampling and non-sampling errors is important for all studies using appropriate protocols and procedures for establishing data quality objectives (DQO) and quality control and assurance (QC/QA).

- Stratified random sampling is sometimes useful for ensuring that the units that are measured are well representative of the population. However, there are potential problems due to the criteria used for stratification being incorrect or the wish to use a different form of stratification to analyse the data after they have been collected.

- Equations for estimating the population mean and total are provided for stratified sampling.

- Post-stratification can be used to analyse the results from simple random sampling as if they were obtained from stratified random sampling.

- With systematic sampling the units measured consist of every kth item in a list, or are regularly spaced over the study area. Units may be easier to select than they are with random sampling and estimates may be more precise than they would otherwise be because they represent the whole population well. Treating a systematic sample as a simple random sample may overestimate the true level of sampling error. Two alternatives to this approach (imposing a stratification and joining sample points with a serpentine line) are described.

- Cluster sampling (selecting groups of close sample units for measurement) may be a cost effective alternative to simple random sampling.

- Multi-stage sampling involves randomly sampling large primary units, randomly sampling some smaller units from each selected primary unit, and possibly randomly sampling even smaller units from each secondary unit, etc. This type of sampling plan may be useful when a hierarchic structure already exists in a population.

- Composite sampling is potentially valuable when the cost of measuring sample units is much greater than the cost of selecting and collecting them. It involves mixing several sample units from the same location and then measuring the mixture. If there is a need to identify sample units with extreme values, then special methods are available for doing this without measuring every sample unit.

- Ranked set sampling involves taking a sample of m units and putting them in order from the smallest to largest. The largest unit only is then measured. A second sample of m units is then selected and the second largest unit is measured. This is continued until the mth sample of m units is taken and the smallest unit is measured. The mean of the m measured units will then usually be a more accurate estimate of the population mean than the mean of a simple random sample of m units.

- Ratio estimation can be used to estimate a population mean and total for a variable X when the population mean is known for a second variable U that is approximately proportional to X. Regression estimation can be used instead if X is approximately linearly related to U.

- Double sampling is an alternative to ratio or regression estimation that can be used if the population mean of U is not known, but measuring U on sample units is inexpensive. A large sample is used to estimate the population mean of U and a smaller sample is used to estimate either a ratio or linear relationship between X and U.

- Methods for choosing sample sizes are discussed. Equations are provided for estimating population means, proportions, differences between means, and differences between proportions.

- Unequal probability sampling is briefly discussed, where the sampling process is such that different units in a population have different probabilities of being included in a sample. Horvitz-Thompson estimators of the total population size and the total for a variable Y are stated.

- The US EPA Data Quality Objective (DQO) process is described. This is a formal seven-step mechanism for ensuring that sufficient data are collected to make required decisions with a reasonable probability that these are correct, and that only the minimum amount of necessary data are collected.

CHAPTER 3

Models for Data

3.1 Statistical Models

Many statistical analyses are based on a specific model for a set of data, where this consists of one or more equations that describe the observations in terms of parameters of distributions and random variables. For example, a simple model for the measurement X made by an instrument might be

$$X = \theta + \epsilon,$$

where θ is the true value of what is being measured, and ϵ is a measurement error that is equally likely to be anywhere in the range from -0.05 to +0.05.

In situations where a model is used, an important task for the data analyst is to select a plausible model and to check, as far as possible, that the data are in agreement with this model. This includes both examining the form of the equation assumed, and the distribution or distributions that are assumed for the random variables.

To aid in this type of modelling process there are many standard distributions available, the most important of which are considered in the following two sections of this chapter. In addition, there are some standard types of model that are useful for many sets of data. These are considered in the later sections of this chapter.

3.2 Discrete Statistical Distributions

A discrete distribution is one for which the random variable being considered can only take on certain specific values, rather than any value within some range (Appendix Section A2). By far the most common situation in this respect is where the random variable is a count and the possible values are 0, 1, 2, 3, and so on.

It is conventional to denote a random variable by a capital X and a particular observed value by a lower case x. A discrete distribution is then defined by a list of the possible values x_1, x_2, x_3, ..., for X, and the probabilities $P(x_1)$, $P(x_2)$, $P(x_3)$, ... for these values. Of necessity,

$$P(x_1) + P(x_2) + P(x_3) + ... = 1,$$

i.e., the probabilities must add to 1. Also of necessity, $P(x_i) \geq 0$ for all i, with $P(x_i) = 0$ meaning that the value x_i can never occur. Often there is a specific equation for the probabilities defined by a probability function

$$P(x) = \text{Prob}(X = x),$$

where P(x) is some function of x.

The mean of a random variable is sometimes called the expected value, and is usually denoted either by μ or E(X). It is the sample mean that would be obtained for a very large sample from the distribution, and it is possible to show that this is equal to

$$E(X) = \sum x_iP(x_i) = x_1P(x_1) + x_2P(x_2) + x_3P(x_3) + ... \quad (3.1)$$

The variance of a discrete distribution is equal to the sample variance that would be obtained for a very large sample from the distribution. It is often denoted by σ^2, and it is possible to show that this is equal to

$$\sigma^2 = \sum (x_i - \mu)^2P(x_i)$$

$$= (x_1 - \mu)^2P(x_1) + (x_2 - \mu)^2P(x_2) + (x_3 - \mu)^2P(x_3) + ... \quad (3.2)$$

The square root of the variance, σ, is the standard deviation of the distribution.

The following discrete distributions are the ones which occur most often in environmental and other applications of statistics. Johnson and Kotz (1969) provide comprehensive details on these and many other discrete distributions.

The Hypergeometric Distribution

The hypergeometric distribution arises when a random sample of size n is taken from a population of N units. If the population contains R units with a certain characteristic, then the probability that the sample will contain exactly x units with the characteristic is

$$P(x) = {}^RC_x\ {}^{N-R}C_{n-x} / {}^NC_n, \text{ for } x = 0, 1, ..., \text{Min}(n,R), \quad (3.3)$$

where ${}^aC_b$ denotes the number of combinations of a objects taken b at at time. The proof of this result will be found in many elementary statistics texts. A random variable with the probabilities of different values given by equation (3.3) is said to have a hypergeometric distribution. The mean and variance are

$$\mu = nR/N, \tag{3.4}$$

and

$$\sigma^2 = R(N - R)n/N^2. \tag{3.5}$$

As an example of a situation where this distribution applies, suppose that a grid is set up over a study area and the intersection of the horizontal and vertical grid lines defines N possible sample locations. Let R of these locations have values in excess of a constant C. If a simple random sample of n of the N locations is taken, then equation (3.1) gives the probability that exactly x out of the n sampled locations will have a value exceeding C.

Figure 3.1(a) shows examples of probabilities calculated for some particular hypergeometric distributions.

The Binomial Distribution

Suppose that it is possible to carry out a certain type of trial and when this is done the probability of observing a positive result is always p for each trial, irrespective of the outcome of any other trial. Then if n trials are carried out the probability of observing exactly x positive is given by the binomial distribution

$$P(x) = {}^nC_x p^x(1 - p)^{n-x}, \text{ for } x = 0, 1, 2, ..., n, \tag{3.6}$$

which is a result also provided in Section A2 of Appendix A. The mean and variance of this distribution are

$$\mu = np, \tag{3.7}$$

and

$$\sigma^2 = np(1 - p), \tag{3.8}$$

respectively.

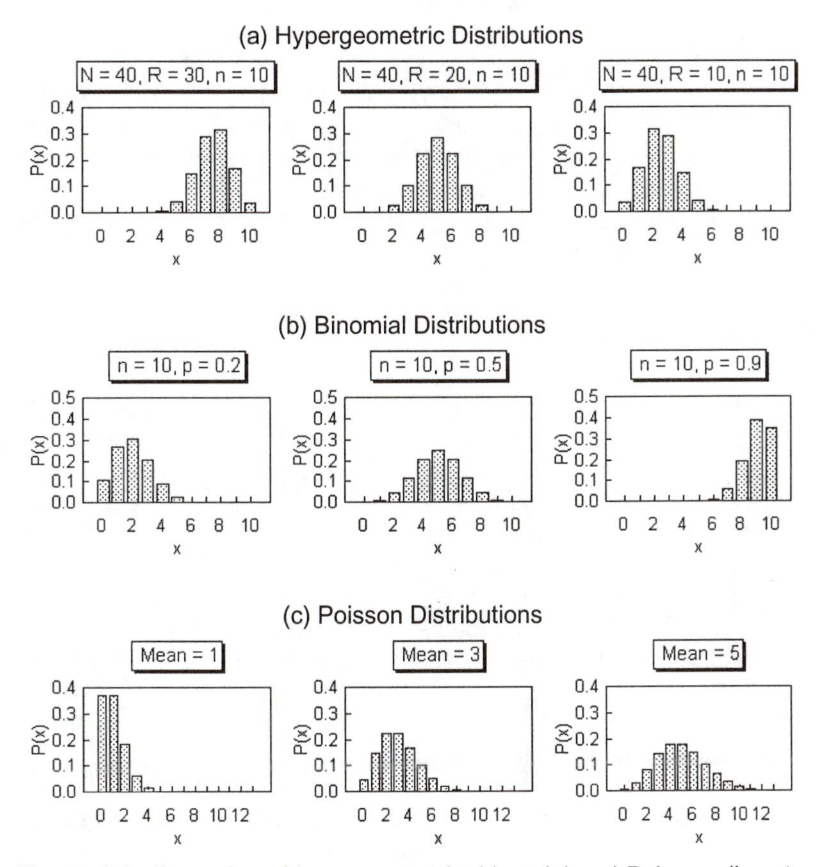

Figure 3.1 Examples of hypergeometric, binomial and Poisson discrete probability distributions.

An example of this distribution occurs with the situation described in Example 1.3, which was concerned with the use of mark-recapture methods to estimate survival rates of salmon in the Snake and Columbia Rivers in the Pacific Northwest of the United States. In that setting, if n fish are tagged and released into a river and there is a probability p of being recorded while passing a detection station downstream for each of the fish, then the probability of recording a total of exactly p fish downstream is given by equation (3.6).

Figure 3.1(b) shows some examples of probabilities calculated for some particular binomial distributions.

The Poisson Distribution

One derivation of the Poisson distribution is as the limiting form of the binomial distribution as n tends to infinity and p tends to zero, with the mean $\mu = np$ remaining constant. More generally, however, it is possible to derive it as the distribution of the number of events in a given interval of time or a given area of space when the events occur at random, independently of each other at a constant mean rate. The probability function is

$$P(x) = \exp(-\mu)\, \mu^x / x!, \text{ for } x = 0, 1, 2, \ldots \quad (3.9)$$

The mean and variance are both equal to μ.

In terms of events occurring in time, the type of situation where a Poisson distribution might occur is for counts of the number of occurrences of minor oil leakages in a region per month, or the number of cases per year of a rare disease in the same region. For events occurring in space a Poisson distribution might occur for the number of rare plants found in randomly selected metre square quadrats taken from a large area. In reality, though, counts of these types often display more variation than is expected for the Poisson distribution because of some clustering of the events. Indeed, the ratio of the variance of sample counts to the mean of the same counts, which should be close to one for a Poisson distribution, is sometimes used as an index of the extent to which events do not occur independently of each other.

Figure 3.1(c) shows some examples of probabilities calculated for some particular Poisson distributions.

3.3 Continuous Statistical Distributions

Continuous distributions are often defined in terms of a probability density function, f(x), which is a function such that the area under the plotted curve between two limits a and b gives the probability of an observation within this range, as shown in Figure 3.2. This area is also the integral between a and b, so that in the usual notation of calculus

$$\text{Prob}(a < X < b) = \int_a^b f(x)\, dx. \quad (3.10)$$

The total area under the curve must be exactly one, and f(x) must be greater than or equal to zero over the range of possible values of x for the distribution to make sense.

The mean and variance of a continuous distribution are the sample mean and variance that would be obtained for a very large random sample from the distribution. In calculus notation the mean is

$$\mu = \int x\, f(x)\, dx,$$

where the range of integration is the possible values for the x. This is also sometimes called the expected value of the random variable X, and denoted E(X). Similarly, the variance is

$$\sigma^2 = \int (x - \mu)^2 f(x)\, dx, \tag{3.11}$$

where again the integration is over the possible values of x.

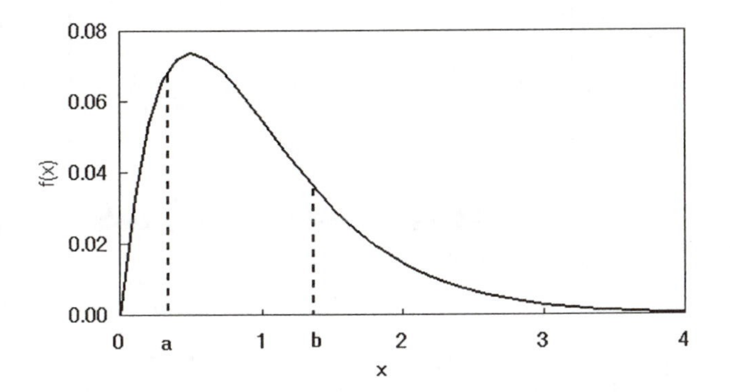

Figure 3.2 The probability density function f(x) for a continuous distribution. The probability of a value between a and b is the area under the curve between these values, i.e., the area between the two vertical lines at x = a and x = b.

The continuous distributions that are described here are ones that often occur in environmental and other applications of statistics. See Johnson and Kotz (1970a, 1970b) for details about many more continuous distributions.

The Exponential Distribution

The probability density function for the exponential distribution with mean μ is

$$f(x) = (1/\mu)\exp(-x/\mu), \text{ for } x \geq 0, \quad (3.12)$$

which has the form shown in Figure 3.3. For this distribution the standard deviation is always equal to the mean μ.

The main application is as a model for the time until a certain event occurs, such as the failure time of an item being tested, the time between the reporting of cases of a rare disease, etc.

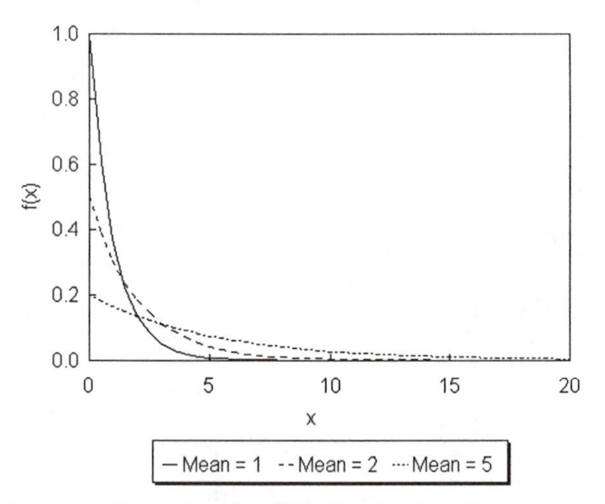

Figure 3.3 Examples of probability density functions for exponential distributions.

The Normal or Gaussian Distribution

The normal or Gaussian distribution with a mean of μ and a standard deviation of σ has the probability density function

$$f(x) = \{1/\sqrt{(2\pi\sigma^2)}\} \exp\{-(x - \mu)^2/(2\sigma^2)\}, \text{ for } -\infty < x < +\infty. \quad (3.13)$$

This distribution is discussed in Section A2 of Appendix A, and the form of the probability density function is illustrated in Figure A1.

The normal distribution is the 'default' that is often assumed for a distribution that is known to have a symmetric bell-shaped form, at least roughly. It is often observed for biological measurements such as the height of humans, and it can be shown theoretically (through something called the central limit theorem) that the normal distribution will tend to result whenever the variable being considered consists of a sum of contributions from a number of other distributions. In particular, mean values, totals, and proportions from simple random samples will often be approximately normally distributed, which is the basis for the approximate confidence intervals for population parameters that have been described in Chapter 2.

The Lognormal Distribution

It is a characteristic of the distribution of many environmental variables that they are not symmetric like the normal distribution. Instead, there are many fairly small values and occasional extremely large values. This can be seen, for example, in the measurements of PCB concentrations that are shown in Table 2.3.

With many measurements only positive values can occur, and it turns out that the logarithm of the measurements has a normal distribution, at least approximately. In that case the distribution of the original measurements can be assumed to be a lognormal distribution, with probability density function

$$f(x) = [1/\{x\sqrt{(2\pi\sigma^2)}\}]\exp[-\{\log_e(x) - \mu\}^2/\{2\sigma^2\}], \text{ for } x > 0. \quad (3.14)$$

Here μ and σ are the mean and standard deviation of the natural logarithm of the original measurement. The mean and variance of the original measurement itself are

$$E(X) = \exp(\mu + \tfrac{1}{2}\sigma^2) \quad (3.15)$$

and

$$Var(X) = \exp(2\mu + \sigma^2)\{\exp(\sigma^2) - 1\}. \quad (3.16)$$

Figure 3.4 shows some examples of probability density functions for three lognormal distributions.

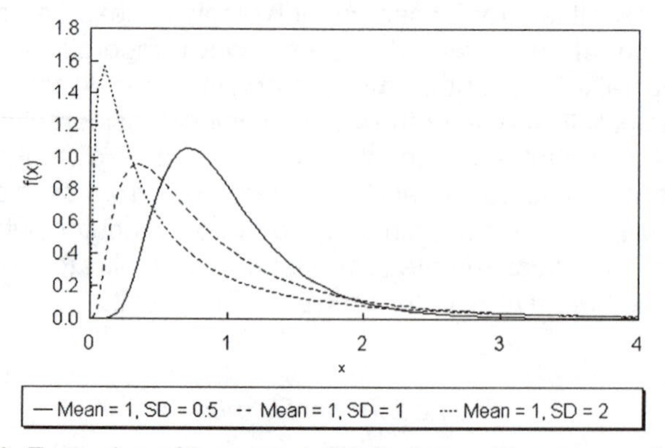

Figure 3.4 Examples of lognormal distributions with a mean of 1.0. The standard deviations are 0.5, 1.0 and 2.0.

3.4 The Linear Regression Model

Linear regression is one of the most frequently used statistical tools. Its purpose is to relate the values of a single variable Y to one or more other variables $X_1, X_2, ..., X_p$, in an attempt to account for the variation in Y in terms of variation in the other variables. With only one other variable this is often referred to as simple linear regression.

The usual situation is that the data available consist of n observations $y_1, y_2, ..., y_n$ for the dependent variable Y, with corresponding values for the X variables. The model that is assumed is

$$y = \beta_0 + \beta_1 x_1 + \beta_2 x_2 + ... + \beta_p x_p + \epsilon, \quad (3.17)$$

where ϵ is a random error with a mean of zero and a constant standard deviation σ. The model is estimated by finding the coefficients of the X values that make the error sum of squares as small as possible. In other words, if the estimated equation is

$$\hat{y} = b_0 + b_1 x_1 + b_2 x_2 + ... + b_p x_p, \quad (3.18)$$

then the b values are chosen so as to minimise

$$SSE = \Sigma(y_i - \hat{y}_i)^2, \quad (3.19)$$

where the $\hat{y}_i$ is the value given by the fitted equation that corresponds to the data value y_i, and the sum is over the n data values. Statistical packages or spreadsheets are readily available to do these calculations.

There are various ways that the usefulness of a fitted regression equation can be assessed. One involves partitioning the variation observed in the Y values into parts that can be accounted for by the X values, and a part (SSE, above) which cannot be accounted for. To this end, the total variation in the Y values is measured by the total sum of squares

$$SST = \Sigma(y_i - \bar{y})^2. \qquad (3.20)$$

This is partitioned into the sum of squares for error (SSE), and the sum of squares accounted for by the regression (SSR), so that

$$SST = SSR + SSE.$$

The proportion of the variation in Y accounted for by the regression equation is then the coefficient of multiple determination,

$$R^2 = SSR/SST = 1 - SSE/SST, \qquad (3.21)$$

which is a good indication of the effectiveness of the regression.

There are a variety of inference procedures that can be applied in the multiple regression situation when the regression errors ϵ are assumed to be independent random variables from a normal distribution with a mean of zero and constant variance σ^2. A test for whether the fitted equation accounts for a significant proportion of the total variation in Y can be based on Table 3.1, which is a variety of what is called an 'analysis of variance table' because it compares the observed variation in Y accounted for by the fitted equation with the variation due to random errors. From this table, the F-ratio,

$$F = MSR/MSE = [SSR/p]/[SSE/(n - p - 1)] \qquad (3.22)$$

can be tested against the F-distribution with p and n - p - 1 degrees of freedom to see if it is significantly large. If this is the case, then there is evidence that Y is related to at least one of the X variables.

Table 3.1 Analysis of variance table for a multiple regression analysis

Source of variation	Sum of squares	Degrees of freedom	Mean square	F-ratio
Regression	SSR	p	MSR	MSR/MSE
Error	SSE	n - p - 1	MSE	
Total	SST	n - 1		

The estimated regression coefficients can also be tested individually to see whether they are significantly different from zero. If this is not the case for one of these coefficients, then there is no evidence that Y is related to the X variable concerned. The test for whether β_j is significantly different from zero involves calculating the statistic $b_j/\hat{SE}(b_j)$, where $\hat{SE}(b_j)$ is the estimated standard error of b_j, which should be output by the computer program used to fit the regression equation. This statistic can then be compared with the percentage points of the t-distribution with n - p - 1 degrees of freedom. If $b_j/\hat{SE}(b_j)$ is significantly different from zero, then there is evidence that β_j is not equal to zero. In addition, if the accuracy of the estimate b_j is to be assessed, then this can be done by calculating a 95% confidence interval for β_j as $b_j \pm t_{5\%,n-p-1}\, b_j/\hat{SE}(b_j)$, where $t_{5\%,n-p-1}$ is the absolute value that is exceeded with probability 0.05 for the t-distribution with n - p - 1 degrees of freedom.

There is sometimes value in considering the variation in Y that is accounted for by a variable X_j when this is included in the regression after some of the other variables are already in. Thus if the variables X_1 to X_p are in the order of their importance, then it is useful to successively fit regressions relating Y to X_1, Y to X_1 and X_2, and so on up to Y related to all the X variables. The variation in Y accounted for by X_j after allowing for the effects of the variables X_1 to X_{j-1} is then given by the extra sum of squares accounted for by adding X_j to the model.

To be more precise, let $SSR(X_1,X_2,...,X_j)$ denote the regression sum of squares with variables X_1 to X_j in the equation. Then the extra sum of squares accounted for by X_j on top of X_1 to X_{j-1} is

$$SSR(X_j|X_1,X_2,...,X_{j-1}) = SSR(X_1,X_2,...,X_j) - SSR(X_1,X_2,...,X_{j-1}). \quad (3.23)$$

On this basis, the sequential sums of squares shown in Table 3.2 can be calculated. In this table the mean squares are the sums of squares divided by their degrees of freedom, and the F-ratios are the mean squares divided by the error mean square. A test for the variable X_j being significantly related to Y, after allowing for the effects of the variables X_1 to X_{j-1}, involves seeing whether the corresponding F-ratio is significantly large in comparison to the F-distribution with 1 and n - p - 1 degrees of freedom.

Table 3.2 Analysis of variance table for the extra sums of squares accounted for by variables as they are added into a multiple regression model one by one

Source of variation	Sum of squares	Degrees of freedom	Mean square	F-ratio
X_1	$SSR(X_1)$	1	$MSR(X_1)$	$F(X_1)$
$X_2 \mid X_1$	$SSR(X_2 \mid X_1)$	1	$MSR(X_2 \mid X_1)$	$F(X_2 \mid X_1)$
.	.	.		
.	.	.		
$X_p \mid X_1,...X_{p-1}$	$SSR(X_p \mid X_1,...X_{p-1})$	1	$MSR(X_p \mid X_1,...X_{p-1})$	$F(X_p \mid X_1,...X_{p-1})$
Error	SSE	n - p - 1	MSE	
Total	SST	n - 1		

If the X variables are uncorrelated, then the F ratios indicated in Table 3.2 will be the same irrespective of what order the variables are entered into the regression. However, usually the X variables are correlated and the order may be of crucial importance. This merely reflects the fact that with correlated X variables it is generally only possible to talk about the relationship between Y and X_j in terms of which of the other X variables are controlled for.

This has been a very brief introduction to the uses of multiple regression. It is a tool that is used for a number of applications later in this book. For a more detailed discussion see Manly (1992, Chapter 4), or one of the many books devoted to this topic (e.g., Neter *et al.*, 1983 or Younger, 1985). Some further aspects of the use of this method are also considered in the following example.

Example 3.1 Chlorophyll-a in Lakes

The data for this example are part of a larger data set originally published by Smith and Shapiro (1981), and also discussed by Dominici *et al.* (1997). The original data set contains 74 cases, where each case consists of observations on the concentration of chlorophyll-a, phosphorus, and (in most cases) nitrogen at a lake at a certain time. For the present example, 25 of the cases were randomly selected from those where measurements on all three variables are present. This resulted in the values shown in Table 3.3.

Chlorophyll-a is a widely used indicator of lake water quality. It is a measure of the density of algal cells, and reflects the clarity of the water in a lake. High concentrations of chlorophyll-a are associated with high algal densities and poor water quality, a condition known as eutrophication. Phosphorus and nitrogen stimulate algal growth and high values for these chemicals are therefore expected to be associated with high chlorophyll-a. The purpose of this example is to illustrate the use of multiple regression to obtain an equation relating chlorophyll-a to the other two variables.

The regression equation

$$CH = \beta_0 + \beta_1 PH + \beta_2 NT + \epsilon \qquad (3.24)$$

was fitted to the data in Table 3.3, where CH denotes chlorophyll-a, PH denotes phosphorus, and NT denotes nitrogen. This gave

$$CH = -9.386 + 0.333PH + 1.200NT, \qquad (3.25)$$

with an R^2 value from equation (3.21) of 0.774. The equation was fitted using the regression option in a spreadsheet, which also provided estimated standard errors for the coefficients of $\hat{SE}(b_1) = 0.046$ and $\hat{SE}(b_2) = 1.172$.

Table 3.3 Values of chlorophyll-a, phosphorus and nitrogen taken from various lakes at various times

Case	Chlorophyll-a	Phosphorus	Nitrogen
1	95.0	329.0	8
2	39.0	211.0	6
3	27.0	108.0	11
4	12.9	20.7	16
5	34.8	60.2	9
6	14.9	26.3	17
7	157.0	596.0	4
8	5.1	39.0	13
9	10.6	42.0	11
10	96.0	99.0	16
11	7.2	13.1	25
12	130.0	267.0	17
13	4.7	14.9	18
14	138.0	217.0	11
15	24.8	49.3	12
16	50.0	138.0	10
17	12.7	21.1	22
18	7.4	25.0	16
19	8.6	42.0	10
20	94.0	207.0	11
21	3.9	10.5	25
22	5.0	25.0	22
23	129.0	373.0	8
24	86.0	220.0	12
25	64.0	67.0	19

To test for the significance of the estimated coefficients, the ratios

$$b_1/\hat{SE}(b_1) = 0.333/0.046 = 7.21,$$

and

$$b_2/\hat{SE}(b_2) = 1.200/1.172 = 1.02$$

must be compared with the t-distribution with $n - p - 1 = 25 - 2 - 1 = 22$ degrees of freedom. The probability of obtaining a value as far from zero as 7.21 is 0.000 to three decimal places, so that there is very strong evidence that chlorophyll-a is related to phosphorus. However, the probability of obtaining a value as far from zero as 1.02 is 0.317, which is quite large. Therefore there seems to be little evidence that chlorophyll-a is related to nitrogen.

This analysis seems straightforward but there are in fact some problems with it. These problems are indicated by plots of the regression residuals, which are the differences between the observed concentrations of chlorophyll-a and the amounts that are predicted by the fitted equation (3.25). To show this it is convenient to use standardized residuals, which are the differences between the observed CH values and the values predicted from the regression equation, divided by the estimated standard deviation of the regression errors.

For a well-fitting model these standardized residuals will appear to be completely random, and should be mostly within the range from -2 to +2. No patterns should be apparent when they are plotted against the values predicted by the regression equation, or the variables being used to predict the dependent variable. This is because the standardized residuals should approximately equal the error term ϵ in the regression model but scaled to have a standard deviation of one.

The standardized residuals are plotted on the left-hand side of Figure 3.5 for the regression equation (3.25). There is some suggestion that (i) the variation in the residuals increases with the fitted value, or, at any rate, is relatively low for the smallest fitted values, (ii) all the residuals are less than zero for lakes with very low phosphorus concentrations, and (iii) the residuals are low, then tend to be high, and then tend to be low again as the nitrogen concentration increases.

The problem here seems to be the particular form assumed for the relationship between chlorophyll-a and the other two variables. It is more usual to assume a linear relationship in terms of logarithms, i.e.,

$$\log(CH) = \beta_0 + \beta_1\log(PH) + \beta_2\log(NT) + \epsilon, \qquad (3.26)$$

for the variables being considered (Dominici *et al.*, 1997). Using logarithms to base ten, fitting this equation by multiple regression gives

$$\log(CH) = -1.860 + 1.238\log(PH) + 0.907\log(NT). \qquad (3.27)$$

The R^2 value from equation (3.21) is 0.878, which is substantially higher than the value of 0.774 found from fitting equation (3.25). The estimated standard errors for the estimated coefficients of log(PH) and log(NT) are 0.124 and 0.326, which means that there is strong evidence that log(CH) is related to both of these variables (t =

1.238/0.124 = 9.99 for log(CH), giving p = 0.000 for the t-test with 22 degrees of freedom; t = 0.970/0.326 = 2.78 for log(NT), giving p = 0.011 for the t-test). Finally, the plots of standardized residuals for equation (3.27) that are shown on the right-hand side of Figure 3.5 give little cause for concern.

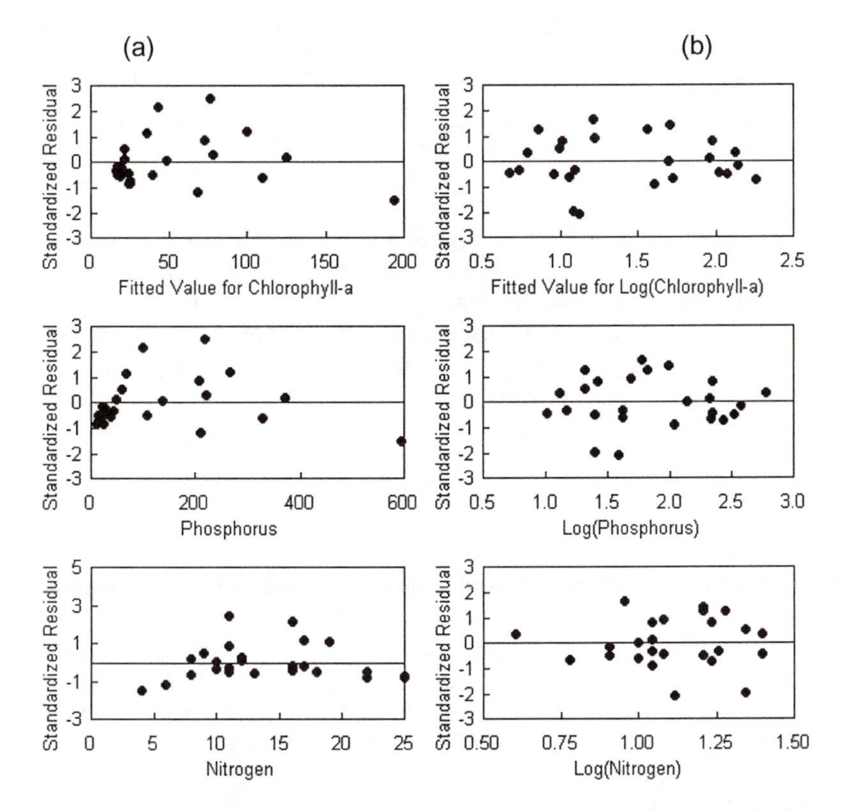

Figure 3.5 (a) Standardized residuals for chlorophyll-a plotted against the fitted value predicted from the regression equation (3.25) and against the phosphorus and nitrogen concentrations for lakes, and (b) standardized residuals for log(chlorophyll-a) plotted against the fitted value, log(phosphorus), and log(nitrogen) for the regression equation (3.27).

An analysis of variance is provided for equation (3.27) in Table 3.4. This shows that the equation with log(PH) included accounts for a very highly significant part of the variation in log(CH). Adding in log(NT) to the equation then gives a highly significant improvement.

Table 3.4 Analysis of variance for equation (3.27) showing the sums of squares accounted for by log(PH), and log(NT) added into the equation after log(PH)

Source	Sum of Squares	Degrees of Freedom	Mean Square	F	p-value
Phosphorus	5.924	1	5.924	150.98	0.0000
Nitrogen	0.303	1	0.303	7.72	0.0110
Error	0.863	22	0.039		
Total	7.090	24	0.295		

In summary, a simple linear regression of chlorophyll-a against phosphorus and nitrogen does not seem to fit the data altogether properly, although it accounts for about 77% of the variation in chlorophyll-a. However, by taking logarithms of all the variables a fit with better properties is obtained, which accounts for about 88% of the variation in log(chlorophyll-a).

3.5 Factorial Analysis of Variance

The analysis of variance that can be carried out with linear regression is very often used in other situations as well, particularly with what are called factorial experiments. An important distinction in this connection is between variables and factors. A variable is something like the phosphorus concentration or nitrogen concentration in lakes, as in the example just considered. A factor, on the other hand, has a number of levels and in terms of a regression model it may be thought plausible that the response variable being considered has a mean level that changes with these levels.

Thus if an experiment is carried out to assess the effect on the survival time of fish of a toxic chemical, then the survival time might be related by a regression model to the dose of the chemical, perhaps at four concentrations, which would then be treated as a variable. If the experiment was carried out on fish from three sources, or on three different species of fish, then the type of fish would be a factor, which could not just be entered as a variable. The fish types would be labelled 1 to 5 and what would be required in the regression equation is that the mean survival time varied with the type of fish.

The type of regression model that could then be considered would be

$$Y = \beta_1X_1 + \beta_2X_2 + \beta_3X_3 + \beta_4X_4 + \epsilon, \qquad (3.28)$$

where Y is the survival time of a fish, X_i for i = 1 to 3 are dummy indicator variables such that $X_i = 1$ if the fish is of type i, or is otherwise 0, and X_4 is the concentration of the chemical. The effect of this formulation is that for a fish of type 1 the expected survival time with a concentration of X_4 is $\beta_1 + \beta_4X_4$, for a fish of type 2 the expected survival time with this concentration is $\beta_2 + \beta_4X_4$, and for a fish of type 3 the expected survival time with this concentration is $\beta_3 + \beta_4X_4$. Hence in this situation the fish type factor at three levels can be allowed for by introducing three 0-1 variables into the regression equation and omitting the constant term β_0.

Equation (3.28) allows for a factor effect, but only on the expected survival time. If the effect of the concentration of the toxic chemical may also vary with the type of fish, then the model can be extended to allow for this, by adding products of the 0-1 variables for the fish type with the concentration variable to give

$$Y = \beta_1X_1 + \beta_2X_2 + \beta_3X_3 + \beta_4X_1X_4 + \beta_5X_2X_4 + \beta_6X_3X_4 + \epsilon. \quad (3.29)$$

For fish of types 1 to 3 the expected survival times are then $\beta_1 + \beta_4X_4$, $\beta_2 + \beta_5X_4$, and $\beta_3 + \beta_6X_4$, respectively. The effect is then a linear relationship between the survival time and the concentration of the chemical which differs for the three types of fish.

When there is only one factor to be considered in a model it can be handled reasonably easily by using dummy indicator variables as just described. However, with more than one factor this gets cumbersome and it is more usual to approach modelling from the point of view of a factorial analysis of variance. This is based on a number of standard models and the theory can get quite complicated. Nevertheless, the use of analysis of variance in practice can be quite straightforward if a statistical package is available to do the calculations. An introduction to experimental designs and their corresponding analyses of variance is given by Manly (1992, Chapter 7), and a more detailed account by Mead *et al.* (1993). Here only three simple situations will be considered.

One Factor Analysis of Variance

With a single factor the analysis of variance model is just a model for comparing the means of I samples, where I is two or more. This model can be written as

$$x_{ij} = \mu + a_i + \epsilon_{ij}, \tag{3.30}$$

where x_{ij} is the jth observed value of the variable of interest at the ith factor level (i.e., in the ith sample), μ is an overall mean level, a_i is the deviation from μ for the ith factor level with $a_1 + a_2 + ... a_I = 0$, and ϵ_{ij} is the random component of xij, which is assumed to be independent of all other terms in the model, with a mean of zero and a constant variance.

To test for an effect of the factor an analysis of variance table is set up, which takes the form shown in Table 3.5. Here the sum of squares for the factor is just the sum of squares accounted for by allowing the mean level to change with the factor level in a regression model, although it is usually computed somewhat differently. The F-test requires the assumption that the random components ϵ_{ij} in the model (3.30) have a normal distribution.

Table 3.5 Form of the analysis of variance table for a one factor model, with I levels of the factor and n observations in total

Source of variation	Sum of Squares[1]	Degrees of freedom	Mean square[2]	F[3]
Factor	SSF	I - 1	MSF = SSB/(I - 1)	MSF/MSE
Error	SSE	n - I	MSE = SSE/(n - I)	
Total	$SST = \sum\sum(x_{ij} - \bar{x})^2$	n - 1		

[1]SSF = sum of squares between factor levels, SSE = sum of squares for error (variation within factor levels), and SST = total sum of squares for which the summation is over all observations at all factor levels.
[2]MSF= mean square between factor levels, and MSE = mean square error.
[3]The F-value is tested for significance by comparison with critical values for the F-distribution with I - 1 and n - I degrees of freedom.

Two Factor Analysis of Variance

With a two factor situation there are I levels for one factor (A) and J levels for the other factor (B). It is simplest if m observations are taken for each combination of levels, which is what will be assumed here. The model can be written

$$x_{ijk} = \mu + a_i + b_j + (ab)_{ij} + \epsilon_{ijk}, \qquad (3.31)$$

where x_{ijk} denotes the kth observation at the ith level for factor A and the jth level for factor B, μ denotes an overall mean level, a_i denotes an effect associated with the ith level of factor A, b_j denotes an effect associated with the jth level of factor B, $(ab)_{ij}$ denotes an interaction effect so that the mean level at a factor combination does not have to be just the sum of the effects of the two individual factors, and ϵ_{ijk} is the random part of the observation x_{ijk}, which is assumed to be independent of all other terms in the model, with a mean of zero and a constant variance.

Moving from one to two factors introduces the complication of deciding whether the factors have what are called fixed or random effects, because this can affect the conclusions reached. With a fixed effects factor the levels of the factor for which data are collected are regarded as all the levels of interest. The effects associated with that factor are then defined to add to zero. Thus if A has fixed effects, then $a_1 + a_2 + ... + a_I = 0$ and $(ab)_{1j} + (ab)_{2j} + ... + (ab)_{Ij} = 0$, for all j. If, on the contrary, A has random effects, then the values a_1 to a_I are assumed to be random values from a distribution with mean zero and variance σ^2_A, while $(ab)_{1j}$ to $(ab)_{Ij}$ are assumed to be random values from a distribution with mean zero and variance σ^2_{AB}.

An example of a fixed effect is when an experiment is run with low, medium and high levels for the amount of a chemical because in such a case the levels can hardly be thought of as a random choice from a population of possible levels. An example of a random effect is when one of the factors in an experiment is the brood of animals tested, where these broods are randomly chosen from a large population of possible broods. In this case the brood effects observed in the data will be random values from the distribution of brood effects that are possible.

The distinction between fixed and random effects is important because the way that the significance of factor effects is determined depends on what is assumed about these effects. Some statistical packages allow the user to choose which effects are fixed and which are random, and carries out tests based on this choice. The 'default' is usually fixed effects for all factors, in which case the analysis of variance table is as shown in Table 3.6.

If there is only m = 1 observation for each factor combination, then the error sum of squares shown in Table 3.6 cannot be calculated. In that case it is usual to assume that there is no interaction between the two factors, in which case the interaction sum of squares becomes an error sum of squares, and the factor effects are tested using F-ratios that are the factor mean squares divided by this error sum of squares.

Table 3.6 Form of the analysis of variance table for a two factor model with fixed effects, and with I levels for factor A, J levels for factor B, m observations for each combination of factor levels, and n = IJm observations in total

Source of variation	Sum of Squares[1]	Degrees of freedom	Mean square	F^2
Factor A	SSA	I - 1	MSA = SSA/(I - 1)	MSA/MSE
Factor B	SSB	J - 1	MSB = SSB/(J - 1)	MSB/MSE
Interaction	SSAB	(I - 1)(J - 1)	MSAB = SSAB/{(I - 1)(J - 1)}	MSAB/MSE
Error	SSE	IJ(m - 1)	MSE = SSE/{IJ(m - 1)}	
Total	$SST = \sum\sum\sum(x_{ijk} - \bar{x})^2$	n - 1		

[1]The sum for SST is over all levels for i, j and k, i.e., over all n observations.
[2]The F-ratios for the factors are for fixed effects only.

Three Factor Analysis of Variance

With three factors with levels I, J, and K, and m observations for each factor combination, the analysis of variance model becomes

$$x_{ijku} = a_i + b_j + c_k + (ab)_{ij} + (ac)_{ik} + (bc)_{jk} + (abc)_{ijk} + \epsilon_{ijku}, \quad (3.32)$$

where x_{ijku} is the uth observation for level i of factor A, level j of factor B, and level k of factor C, a_i, b_j and c_k are the main effects of the three factors, $(ab)_{ij}$, $(ac)_{ik}$ and $(bc)_{jk}$ are terms that allow for first order interactions between pairs of factors, $(abc)_{ijk}$ allows for a three factor interaction (where the mean for a factor combination is not just the sum of the factor and first order interaction effects), and ϵ_{ijku} is a random component of the observation, independent of all other terms in the model with a mean of zero and a constant variance.

The analysis of variance table generalises in an obvious way in moving from two to three factors. There are now sums of squares, mean squares and F-ratios for each of the factors, the two factor interactions, the three factor interaction, and the error term, as shown in Table 3.7. This table is for all effects fixed. With one or more random effects some of the F-ratios must be computed differently.

Example 3.2 Survival of Trout in a Metals Mixture

This example concerns part of the results from a series of experiments conducted by Marr *et al.* (1995) to compare the survival of naive and metals-acclimated juvenile brown trout (*Salmo trutta*) and rainbow trout (*Oncorhynchus mykiss*) when exposed to a metals mixture with the maximum concentrations found in the Clark Fork River, Montana, USA.

In the trials called challenge 1 there were three groups of fish (hatchery brown trout, hatchery rainbow trout, and Clark Fork River brown trout). Approximately half of each group (randomly selected) were controls that were kept in clean water for three weeks before being transferred to the metals mixture. The other fish in each group were acclimated for three weeks in a weak solution of metals before being transferred to the stronger mixture. All fish survived the initial three week period, and an outcome variable of interest was the survival time of the fish in the stronger mixture. The results from the trials are shown in Table 3.8.

The results from this experiment can be analysed using the two factor analysis of variance model. The first factor is the type of fish, which is at three levels (two types of brown trout and one type of rainbow trout). This is a fixed effects factor because no other types of fish are being considered. The second factor is the treatment, which is at two levels (control and acclimated). Again this is a fixed effects factor because no other treatments are being considered. A slight complication is the unequal numbers of fish at the different factor combinations. However, many statistical packages can allow for this reasonably easily. The analysis presented here was carried out with the general linear model option in MINITAB (Minitab Inc., 1994).

Table 3.7 Form of the analysis of variance table for a three factor model with fixed effects, and with I levels for factor A, J levels for factor B, K levels for factor C, m observations for each combination of factor levels, and n = IJMm observations in total

Source of variation	Sum of Squares[1]	Degrees of freedom	Mean square	F[2]
Factor A	SSA	I - 1	MSA = SSA/(I - 1)	MSA/MSE
Factor B	SSB	J - 1	MSB = SSB/(J - 1)	MSB/MSE
Factor C	SSC	K - 1	MSC = SSC/(K - 1)	MSC/MSE
AB Interaction	SSAB	(I - 1)(J - 1)	MSAB = SSAB/{(I - 1)((J - 1)}	MSAB/MSE
AC Interaction	SSAC	(I - 1)(K - 1)	MSAC = SSAC/{(I - 1)(K - 1)}	MSAC/MSE
BC Interaction	SSBC	(J - 1)(K - 1)	MSBC = SSBC/{(J - 1)(K - 1)}	MSBC/MSE
ABC Interaction	SSABC	(I - 1)(J - 1)(K - 1)	MSABC = SSABC/{(I - 1)(J - 1)(K - 1)}	MSABC/MSE
Error	SSE	IJK(m - 1)	MSE = SSE/{IJK(m - 1)}	
Total	$SST = \sum\sum\sum\sum(x_{ijk} - \bar{x})^2$	n - 1		

[1]The sum for SST is over all levels for i, j , k and m, i.e., over all n observations.
[2]The F-ratios for the factors and two factor interactions are for fixed effects only.

Table 3.8 Results from Marr *et al.*'s (1995) challenge 1 experiment where the effect of an acclimatization treatment on survival was examined for three types of fish. The tabulated values are survival times in hours

	Hatchery Brown Trout		Hatchery Rainbow Trout		Clark Fork Brown Trout	
	Control	Treated	Control	Treated	Control	Treated
	8	10	24	54	30	36
	18	60	24	48	30	30
	24	60	24	48	30	30
	24	60	24	54	36	30
	24	54	24	54	30	36
	24	72	24	36	36	30
	18	54	24	30	36	42
	18	30	24	18	24	54
	24	36	24	48	36	30
	18	48	24	36	36	48
	10	48	24	24	36	24
	24	42	18	24	30	54
	24	54	18	48	18	54
	24	10	24	48	30	36
	10	66	30	36	24	30
	18	42	30	42	30	90
	24	36	30	36	24	60
	24	42	30	36	30	66
	24	36	36	42	42	108
	24	36	30	36	42	114
	24	36	30	36	24	108
	24	36	30	36	10	114
	24	36	30	36	24	120
	24	36	30	42	24	90
	24	30	36	42	24	96
	24	30	36	36	36	30
	24	36	36	36	24	108
	24	30	36	36	30	108
	24	36	36	36	18	108
	24	36	36		24	102
						102
						120
n	30	30	30	29	30	32
Mean	21.53	41.27	28.20	39.10	28.93	69.00
Std Dev.	4.72	14.33	5.49	8.87	7.23	35.25

A second complication is the increase in the variation in the survival time as the mean increases. It can be seen, for example, that the lowest mean survival time shown in Table 3.8 (21.53 hours) is for control hatchery brown trout. This group also has the lowest standard deviation (4.72 hours). This can be compared with the highest mean survival time (69.00 hours) for acclimated Clark Fork River brown trout, which also has the highest standard deviation (35.25 hours). It seems, therefore, that the assumption of a constant variance for the random component in the model (3.31) is questionable. This problem can be overcome for this example by analysing the logarithm of the survival time rather than the survival time itself. This largely removes the apparent relationship between means and variances.

The analysis of variance is shown in Table 3.9 for logarithms to base 10. Starting from a model with no effects, adding the species factor gives a very highly significant improvement in the fit of the model (F = 17.20, p = 0.000). Adding the main effect of treatment leads to another very highly significant improvement in fit (F = 108.39, p = 0.000). Finally, adding in the interaction gives a highly significant improvement in the fit of the model (F = 5.72, p = 0.004). It can therefore be concluded that the mean value of the logarithm of the survival time varies with the species, and with the acclimation treatment. Also, because of the interaction that seems to be present, the effect of the acclimation treatment is not the same for all three types of fish.

Table 3.9 Analysis of variance on logarithms to base 10 of the daily survival times shown in Table 3.8

Source of Variation	Sum of Squares[1]	Degrees of Freedom	Mean Square	F	p-Value
Species	0.863	2	0.431	17.20	0.000
Treatment	2.719	1	2.719	108.39	0.000
Interaction	0.287	2	0.143	5.72	0.004
Error	4.389	175	0.025		
Total	8.257	180			

[1]The sums of squares shown here depend on the order in which effects are added into the model, which is species, then the treatment, and finally the interaction between these two factors.

On a logarithmic scale, a treatment has no interaction when the proportional change that it causes is constant. For the challenge 1 trials the interaction is significant because the effect of acclimation varies considerably with the three types of fish: for hatchery brown trout the mean survival time for acclimated fish (41.3 hours) was 92% higher than the mean survival time for the controls (21.5 hours); for hatchery rainbow trout the mean survival time for acclimated fish (39.1 hours) was 39% higher than the mean survival time for the controls (28.2 hours); and for Clark Fork River brown trout the mean survival time for the acclimated fish (69.0 hours) was 139% higher than the mean survival time for the controls (28.9 hours). The proportional changes are therefore very far from being constant.

The assumptions of the analysis of variance model seem fairly reasonable for this example after the survival times are transformed to logarithms, as can be seen from the plots of standardized residuals that are shown in Figure 3.6. For a good fitting model most of the standardized residuals should be within the range -2 to +2, which they are. There are, however, some standardized residuals less than -2, and one of nearly -4. There is also a suggestion that the amount of variation is relatively low for fish with a small expected survival time, the second type of fish (hatchery rainbow trout), and treatment 1 (the controls). These effects are not clear enough to cause much concern.

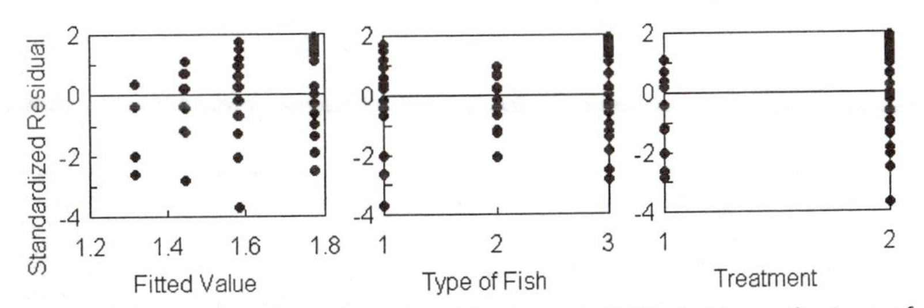

Figure 3.6 Plots of standardized residuals against fitted values, the type of fish, and the treatment from the analysis of variance model for the logarithm of survival times from Marr *et al.*'s (1995) challenge 1 trials.

Repeated Measures Designs

Many environmental data sets have a repeated measures type of design. An example would be vegetation monitoring to assess the effect of alternative control strategies on browsing pests, such as possums in New Zealand. There might, for example, be three areas, one with no pest control, one with some pest control, and one with intensive pest control. Within each area four randomly placed plots might be set up, and then the percentage foliage cover measured for six years on those plots. This would then result in data of the form shown in Table 3.10.

Table 3.10 The form of data from a repeated measures experiment with four plots in each of three different treatment areas measured for six years. A measurement of percentage foliage cover is indicated by X

Area	Plot	Year 1	Year 2	Year 3	Year 4	Year 5	Year 6
No pest control	1	X	X	X	X	X	X
	2	X	X	X	X	X	X
	3	X	X	X	X	X	X
	4	X	X	X	X	X	X
Low pest control	1	X	X	X	X	X	X
	2	X	X	X	X	X	X
	3	X	X	X	X	X	X
	4	X	X	X	X	X	X
High pest control	1	X	X	X	X	X	X
	2	X	X	X	X	X	X
	3	X	X	X	X	X	X
	4	X	X	X	X	X	X

In this example the area is a between-plot factor at three levels, and the year is a within-plot factor. There is a special option in many statistics packages to analyse data of this type, and there can be more than one between plot factor, and more than one within plot factor. A set of data like that shown in Table 3.10 should not be analysed as a factorial design with three factors (area, year and plot), because that assumes that the plots in different areas match up, e.g., plot 1 in areas 1, 2 and 3 have something similar about them, which will generally not be true. Rather, the numbering of plots within areas will be arbitrary, which is sometimes referred to as plots being nested within areas. On the other hand, a repeated measures analysis of variance does assume that the measurements at different times on one plot in one area tend to be similar. Thus it is one way of overcoming the problem of pseudoreplication, which is discussed further in Section 4.8.

An important special application of repeated measures analysis of variance is with the before-after-control-impact (BACI) and other designs that are discussed in Chapter 6. In this case there may be a group of sites that are controls, with no potential impact, and another group of sites that are potentially impacted. Repeated observations are made on the sites at different times over the study period, and at some point in time there is an event at the sites that may be impacted. The question then is whether the event has a detectable effect on the observations at the impact sites.

The analysis of designs with repeated measures can be quite complicated, and it is important to make the right assumptions (Von Ende, 1993). This is one area where expert advice may need to be sought.

Multiple Comparisons and Contrasts

Many statistical packages for analysis of variance allow the user to make comparisons of the mean level of the dependent variable for different factor combinations, with the number of multiple comparisons being allowed for in various ways. Multiple testing methods are discussed further in Section 4.9. Basically, they are a way to ensure that the number of significant results is controlled when a number of tests of significance are carried out at the same time, with all the null hypotheses being true. Or, alternatively, the same methods can be used to ensure that when a number of confidence intervals are calculated at the same time, then they will all contain the true parameter value with a high probability.

There are often many options available with statistical packages, and the help facility with the package should be read carefully before deciding which, if any, of these to use. The use of a Bonferroni correction is one possibility that is straightforward and usually available, although this may not have the power of other methods.

Be warned that some statisticians do not like multiple comparison methods. To quote one leading expert on the design and analysis of experiments (Mead, 1988, p. 310):

> *Although each of these methods for multiple comparisons was developed for a particular, usually very limited, situation, in practice these methods are used very widely with no apparent thought as to their appropriateness. For many experimenters, and even editors of journals, they have become automatic in the less desirable sense of being used as a substitute for thought ... I recommend strongly that multiple comparison*

methods be avoided unless, after some thought and identifying the situation for which the test you are considering was proposed, you decide that the method is exactly appropriate.

He goes on to suggest that simple graphs of means against factor levels will often be much more informative than multiple comparison tests.

On the other hand, Mead (1988) does make use of contrasts for interpreting experimental results, where these are linear combinations of mean values that reflect some aspect of the data that is of particular interest. For example, if observations are available for several years, then one contrast might be the mean value in year 1 compared to the mean value for all other years combined. Alternatively, a set of contrasts might be based on comparing each of the other years with year 1. Statistical packages often offer the possibility of considering either a standard set of contrasts, or contrasts defined by the user.

3.6 Generalized Linear Models

The regression and analysis of variance models described in the previous two sections can be considered to be special cases of a general class of generalized linear models. These were first defined by Nelder and Wedderburn (1972), and used to develop GLIM, a computer program for fitting these models to data (Francis *et al.*, 1993). They include many of the regression types of models that are likely to be of most use for analysing environmental data. A very thorough description of the models and the theory behind them is provided by McCullagh and Nelder (1989).

The characteristic of generalized linear models is that there is a dependent variable Y, which is related to some other variables X_1, X_2, ..., X_p by an equation of the form

$$Y = f(\beta_0 + \beta_1 X_1 + \beta_2 X_2 + ... + \beta_p X_p) + \epsilon, \quad (3.33)$$

where f(x) is one of a number of allowed functions, and ϵ is a random value with a mean of zero from one of a number of allowed distributions. For example, setting f(x) = x and assuming a normal distribution for ϵ just gives the usual multiple regression model of equation (3.17).

Setting f(x) = exp(x) makes the expected value of Y equal to

$$E(Y) = \exp(\beta_0 + \beta_1X_1 + \beta_2X_2 + ... + \beta_pX_p). \quad (3.34)$$

Assuming that Y has a Poisson distribution then gives a log-linear model, which is a popular assumption for analysing count data (Manly, 1992, Section 8.5). The description 'log-linear' comes about because the logarithm of the expected value of Y is a linear combination of the X variables.

Alternatively, setting f(x) = exp(x)/{1 + exp(x)} makes the expected value of Y equal to

$$E(Y) = \exp(\beta_0 + \beta_1X_1 +...+ \beta_pX_p)/[1 + \exp(\beta_0 + \beta_1X_1 +...+ \beta_pX_p)]. \quad (3.35)$$

This is the logistic model for a random variable Y that takes the value 0 (indicating the absence of a event) or 1 (indicating that an event occurs), where the probability of Y = 1 is given as a function of the X variables by the right-hand side of equation (3.35).

There are many other possibilities for modelling within this framework either using the GLIM computer program or one of the other statistical programs that do essentially the same thing. Table 3.11 gives a summary of the models that are allowed in version 4 of GLIM.

Generalized linear models are usually fitted to data using the principle of maximum likelihood, i.e., the unknown parameter values are estimated as those values that make the probability of the observed data as large as possible. This principle is one that is often used in statistics for estimation. Here it is merely noted that the goodness of fit of a model is measured by the deviance, which is (apart from a constant) minus twice the maximized log-likelihood, with associated degrees of freedom equal to the number of observations minus the number of estimated parameters.

With models for count data with Poisson errors the deviance gives a direct measure of absolute goodness of fit. If the deviance is significantly large in comparison with critical values from the chi-squared distribution, then the model is a poor fit to the data. Conversely, a deviance that is not significantly large shows that the model is a good fit. Similarly, with data consisting of proportions with binomial errors the deviance is an absolute measure of the goodness of fit when compared to the chi-squared distribution providing that the numbers of trials that the proportions relate to (n in Table 3.11) are not too small, say generally more than five.

With data from distributions other than the Poisson or binomial, or for binomial data with small numbers of trials the deviance can only be used as a relative measure of goodness of fit. The key result then is that if one model has a deviance of D_1 with v_1 degrees of freedom and another model has a deviance of D_2 with v_2 degrees of freedom, and the first model contains all of the parameters in the second model plus some others, then the first model gives a significantly better fit than the second model if the difference $D_2 - D_1$ is significantly large in comparison with the chi-squared distribution with $v_2 - v_1$ degrees of freedom. Comparing several models in this way is called an analysis of deviance by analogy to the analysis of variance. These tests using deviances are approximate but they should give reasonable results, except perhaps with rather small sets of data.

The individual estimates in a generalized linear model can also be tested to see whether they are significantly different from zero. This just involves comparing the estimate divided by its estimated standard error,

$$z = b/\hat{SE}(b),$$

with critical values for the standard normal distribution. Thus if the absolute value of z exceeds 1.96, then the estimate is significantly different from zero at about the 5% level.

One complication that can occur with some generalized linear models is over-dispersion, with the differences between the observed and fitted values of the dependent variable being much larger than what is expected from the assumed error distribution. There are ways of handling this should it arise. For more information on this and other aspects of these models, see Healy (1988), McCullagh and Nelder (1989), Lindsey (1989), and the very comprehensive GLIM 4 manual (Francis *et al.*, 1993).

Table 3.11 Generalized linear models that can be fitted by version 4 of GLIM (Francis *et al*., 1993)

Model Name	Model Equation	Distributions for Y				
		Normal	Poisson	Binomial	Gamma	Inverse Gaussian
Linear regression	$Y = \sum \beta_i X_i + \epsilon$	default	allowed	doubtful	allowed	allowed
Log-linear	$Y = \exp(\sum \beta_i X_i) + \epsilon$	allowed	default	doubtful	allowed	allowed
Logistic regression	$Y/n = \exp(\sum \beta_i X_i)/\{1+\exp(\sum \beta_i X_i)\} + \epsilon$	doubtful	not possible	default	doubtful	doubtful
Reciprocal	$Y = 1/(\sum \beta_i X_i) + \epsilon$	allowed	allowed	doubtful	default	allowed
Probit	$Y/n = \Phi(\sum \beta_i X_i) + \epsilon$	doubtful	not possible	allowed	doubtful	doubtful
Double exponential	$Y/n = 1 - \exp\{-\exp(\sum \beta_i X_i)\} + \epsilon$	doubtful	not possible	allowed	doubtful	doubtful
Square	$Y = (\sum \beta_i X_i)^2 + \epsilon$	allowed	allowed	doubtful	allowed	allowed
Exponential	$Y = (\sum \beta_i X_i)^{1/a} + b + \epsilon$, a and b constants	allowed	allowed	doubtful	allowed	allowed
Square root inverse	$Y = 1/\{\sum \beta_i X_i\}^{½} + \epsilon$	allowed	allowed	doubtful	allowed	default

Example 3.3 Dolphin Bycatch in Trawl Fisheries

The accidental capture of marine mammals and birds in commercial fishing operations is of considerable concern in many fisheries around the world. This example concerns one such situation, which is of catches of the common dolphin (*Delphinus delphis*) and the bottlenose dolphin (*Tursiops truncatus*) in the Taranaki Bight trawl fishery for jack mackerel (*Trachurus declivis*, *T. novaezalandiae*, and *T. murphyi*) off the west coast of New Zealand.

The New Zealand Ministry of Fisheries puts official observers on about 10% of fishing vessels to monitor dolphin bycatch, and Table 3.12 shows a summary of the data collected by these observers for the six fishing seasons 1989/90 to 1994/95, as originally published by Baird (1996, Table 3). The table shows the number of observed trawls and the number of dolphins accidentally killed categorised by eight conditions for each fishing year: the fishing area (the northern or southern Taranaki Bight), by the gear type (bottom or midwater), by the time (day or night). Excluding five cases where there were no observed trawls, this gives 43 observations on the bycatch under different conditions, in different years. Some results from fitting a generalized linear model to these data are also shown in the last two columns of the table.

The data in Table 3.12 will be used here to examine how the amount of bycatch varies with the factors recorded (year, area, gear type, and time of day). Because the dependent variable (the number of dolphins killed) is a count, it is reasonable to try fitting the data using a log-linear model with Poisson errors. A simple model of that type for the ith count is

$$Y_i = T_i\exp\{\alpha(f_i) + \beta_1X_{i1} + \beta_2X_{i2} + \beta_3X_{i3}\} + \epsilon_i, \qquad (3.36)$$

where T_i is the number of tows involved; $\alpha(f_i)$ depends on the fishing year f_i when the observation was collected in such a way that $\alpha(f_i) = \alpha(1)$ for observations in 1989/90, $\alpha(f_i) = \alpha(2)$ for observations in 1990/91, and so on up to $\alpha(f_i) = \alpha(6)$ for observations in 1994/95; X_{i1} is 0 for North Taranaki and 1 for South Taranaki; X_{i2} is 0 for bottom trawls and 1 for mid-water trawls; and X_{i3} is 0 for day and 1 for night. The fishing year is then being treated as a factor at six levels while the three X variables indicate the absence and presence of different particular conditions. The number of trawls is included as a multiplying factor in equation (3.36) because, other things being equal, the amount of bycatch is expected to be proportional to the number of trawls made. Such a multiplying factor is called an offset in the model.

Table 3.12 Bycatch of dolphins in the Taranaki Bight trawl fishery for jack mackerel on tows officially observed

					Dolphins Killed		
Season	Area	Gear Type	Time	Tows	Data	Fitted	Rate[1]
1989-90	North	Bottom	Day	48	0	0.0	0.1
	North	Bottom	Night	6	0	0.0	0.6
	North	Mid-water	Night	1	0	0.0	3.9
	South	Bottom	Day	139	0	0.6	0.4
	South	Mid-water	Day	6	0	0.2	2.8
	South	Bottom	Night	6	0	0.2	3.6
	South	Mid-water	Night	90	23	21.9	24.4
1990-91	North	Bottom	Day	2	0	0.0	0.0
	South	Bottom	Day	47	0	0.0	0.0
	South	Mid-water	Day	110	0	0.0	0.0
	South	Bottom	Night	12	0	0.0	0.0
	South	Mid-water	Night	73	0	0.0	0.0
1991-92	North	Bottom	Day	101	0	0.4	0.4
	North	Mid-water	Day	4	0	0.1	2.8
	North	Bottom	Night	36	2	1.3	3.6
	North	Mid-water	Night	3	5	0.7	24.3
	South	Bottom	Day	74	1	1.9	2.5
	South	Mid-water	Day	3	0	0.5	17.1
	South	Bottom	Night	7	5	1.5	22.1
	South	Mid-water	Night	15	16	22.6	150.4

[1]Dolphins expected to be captured per 100 tows according to the fitted model.

Table 3.12 (Continued)

Season	Area	Gear Type	Time	Tows	Dolphins Killed Observed	Fitted	Rate[1]
1992-93	North	Bottom	Day	135	0	0.1	0.1
	North	Mid-water	Day	3	0	0.0	0.5
	North	Bottom	Night	22	0	0.1	0.6
	North	Mid-water	Night	16	0	0.7	4.2
	South	Bottom	Day	112	0	0.5	0.4
	South	Bottom	Night	6	0	0.2	3.9
	South	Mid-water	Night	28	9	7.4	26.3
1993-94	North	Bottom	Day	78	0	0.0	0.0
	North	Mid-water	Day	19	0	0.0	0.2
	North	Bottom	Night	13	0	0.0	0.2
	North	Mid-water	Night	28	0	0.4	1.6
	South	Bottom	Day	155	0	0.2	0.2
	South	Mid-water	Day	20	0	0.2	1.1
	South	Bottom	Night	14	0	0.2	1.4
	South	Mid-water	Night	71	8	6.8	9.6
1994-95	North	Bottom	Day	17	0	0.0	0.1
	North	Mid-water	Day	80	0	0.3	0.4
	North	Bottom	Night	9	0	0.0	0.5
	North	Mid-water	Night	74	0	2.5	3.4
	South	Bottom	Day	41	0	0.1	0.4
	South	Mid-water	Day	73	6	1.8	2.4
	South	Bottom	Night	13	0	0.4	3.1
	South	Mid-water	Night	74	15	15.8	21.3

[1]Dolphins expected to be captured per 100 tows according to the fitted model.

The model was fitted using GLIM4 (Francis *et al.*, 1993) to produce the estimates that are shown in Table 3.13. The estimates for the effects of different years are not easy to interpret because their estimated standard errors are quite large. Nevertheless there are significant differences between years as will be seen from the analysis of deviance to be considered shortly. The main thing to notice in this respect is the absence of any recorded bycatch in 1990/91.

Table 3.13 Estimates from fitting a log-linear model to the dolphin bycatch data

Parameter	Estimate	Standard Error
α(1), year effect 1989/90	-7.328	0.590
α(2), year effect 1990/91	-17.520	21.380
α(3), year effect 1991/92	-5.509	0.537
α(4), year effect 1992/93	-7.254	0.612
α(5), year effect 1993/94	-8.260	0.636
α(6), year effect 1994/95	-7.463	0.551
Area effect (south v north)	1.822	0.411
Gear effect (mid-water v bottom)	1.918	0.443
Time effect (night v day)	2.177	0.451

The coefficient of X_1, the area effect, is 1.822, with standard error 0.411. This estimate is very highly significantly different from zero ($z = 1.822/0.411 = 4.44$, $p = 0.000$ compared to the standard normal distribution). The positive coefficient indicates that bycatch was higher in South Taranaki than in North Taranaki. Other things being equal the estimated rate in the south is exp(1.822) = 6.18 times higher than the estimated rate in the north.

The estimated coefficient of X_2, the gear type effect, is 1.918 with standard error 0.443. This is very highly significantly different from zero ($z = 4.33$, $p = 0.000$). The estimated coefficient implies that the bycatch rate is exp(1.918) = 6.81 times higher for mid-water trawls than it is for bottom trawls.

The estimated coefficient for X_3, the time of fishing, is 2.177 with standard error 0.451. This is another highly significant result ($z = 4.82$, $p = 0.000$), implying that, other things being equal, the bycatch rate at night is exp(2.177) = 8.82 higher at night than it is during the day.

Table 3.14 shows the analysis of deviance table obtained by adding effects into the model one at a time. All effects are highly significant in terms of the reduction in the deviance that is obtained by adding them into the model, and the final model gives a good fit to the data (chi-squared = 42.08 with 34 degrees of freedom, p = 0.161).

Table 3.14 Analysis of deviance for the log-linear model fitted to the dolphin bycatch data

			Change	
Effect	Deviance	Degrees of freedom	Deviance	Degrees of freedom
No effects	334.33[1]	42		
			58.48[2]	5
+ Year	275.85[1]	37		
			60.71[2]	1
+ Area	215.14[1]	36		
			139.16[2]	1
+ Gear type	75.98[1]	35		
			33.91[2]	1
+ Time	42.07	34		

[1]Significantly large at the 0.1% level, indicating that the model gives a poor fit to the data.
[2]Significantly large at the 0.1% level, indicating that bycatch is very strongly related to the effect added to the model.

Finally, the last two columns of Table 3.12 show the expected counts of dolphin deaths to compare with the observed counts, and the expected number of deaths per 100 tows. The expected number of deaths per 100 tows is usually fairly low but has the very large value of 150.4 for mid-water tows, at night, in South Taranaki, in 1991/92. In summary, it seems clear that bycatch rates seem to have varied greatly with all of the factors considered in this example.

3.7 Chapter Summary

- Statistical models describe observations on variables in terms of parameters of distributions and the nature of the random variation involved.

- The properties of discrete random variables are briefly described, including definitions of the mean and the variance. The hypergeometric, binomial, and Poisson distributions are described.

- The properties of continuous random variables are briefly described, including definitions of a probability density function, the mean and the variance. The exponential, normal, and lognormal distributions are described.

- The theory of linear regression is summarised for relating the values of a variable Y to the corresponding values of some other variables $X_1, X_2, ..., X_p$. The coefficient of multiple determination, R^2 is defined. Tests for the significance of relationships and the use of analysis of variance with regression are described.

- An example on the prediction of chlorophyll-a concentrations from phosphorus and nitrogen concentrations in lakes is used to illustrate the use of multiple regression.

- The difference between factors and variables is described. The models for one, two and three factor analysis of variance are defined, together with their associated analysis of variance tables.

- A two factor example on the survival time of fish is used to illustrate analysis of variance, where the two factors are the type of fish and the treatment before the fish were kept in a mixture of toxic metals.

- The structure of data for a repeated measures analysis of variance is defined.

- The use of multiple testing methods for comparing means after analysis of variance is briefly discussed, and the fact that these methods are not approved of by all statisticians. The use of contrasts is also discussed briefly.

- The structure of generalized linear models is defined. Tests for goodness of fit, and for the existence of effects based on the comparison of deviances with critical values of the chi-squared distribution are described, as are tests for the significance of the coefficients of individual X variables.

- The use of a generalized linear model is illustrated by an example where the number of dolphins accidentally killed during commercial fishing operations is related to the year of fishing, the type of fishing gear used, and the time of day of fishing. A log-linear model with the number of dolphins killed assumed to have a Poisson distribution is found to give a good fit to the data.

CHAPTER 4

Drawing Conclusions from Data

4.1 Introduction

Statistics is all about drawing conclusions from data, and there has been much of this activity in the earlier chapters of this book. The purpose of the present chapter is to look more closely and critically at the methods that are used for drawing conclusions. Quite a variety of topics are considered, including some which are rather important and yet often receive relatively little attention in statistics texts. These include the difference between observational and experimental studies, the difference between inference based on the random sampling design used to collect data and inference based on the assumption of a particular model for the data, criticisms that have been raised about the excessive use of significance tests, the use of computer-intensive methods of randomization and bootstrapping instead of more conventional methods, the avoidance of pseudo-replication, the use of sampling methods where sample units have different probabilities of selection, the problem of multiple testing, meta-analysis (methods for combining the results from different studies), and the use of Bayesian inference, which is currently receiving a great deal of attention.

4.2 Observational and Experimental Studies

When considering the nature of empirical studies there is an important distinction between observational and experimental studies. With observational studies data are collected by observing populations in a passive manner that as much as possible will not change the ongoing processes. For example, samples of animals might be collected to estimate the proportions in different age classes or the sex ratio. On the other hand, experimental studies are usually thought of as involving the collection of data with some manipulation of variables that is assumed to affect population parameters, keeping other variables constant as far as possible. An example of this type would be a study where some animals are removed from an area to see whether they are replaced by invaders from outside.

In many cases the same statistical analysis can be used with either observational or experimental data. However, the validity of any inferences that result from the analysis depends very much on the type of study. In particular, an effect that is seen consistently in replications of a well designed experiment can only reasonably be explained as being caused by the manipulation of the experimental variables. But with an observational study the same consistency of results might be obtained because all the data are affected in the same way by some unknown and unmeasured variable. Therefore, the 'obvious' explanation for an effect that is seen in the results of an observational study may be quite wrong. To put it another way, the conclusions from observational studies are not necessarily wrong. The problem is that there is little assurance that they are right (Hairston, 1989, p. 1).

It is clear that in general it is best to base inferences on experiments rather than observational studies, but this is not always possible. Some experiments cannot be performed either because the variables involved are not controllable, or because the experiment is simply not feasible. For example, suppose that a researcher wishes to assess the effect of discharges of pollutants from a sewage treatment plant on the organisms in a river. Then systematically changing the levels of pollutants, and in some cases increasing them to higher levels than normally occur, might either not be possible or be considered to be unethical. Hence, in this situation the only study possible might be one involving attempts to relate measurements on the organisms to unplanned variation in pollutant levels, with some allowance for the effects of other factors that may be important.

Having defined two categories of study (observational and experimental), it must be admitted that at times the distinction becomes a little blurred. In particular, suppose that the variables thought to determine the state of an ecological system are abruptly changed either by some naturally occurring accident, or are an unintentional result of some human intervention. If the outcome is then studied this appears to be virtually the same as if the changes were made by the observer as part of an experiment. But such 'natural experiments' do not have some of the important characteristics of true experiments. The conclusions that can be drawn might be stronger than those that could be drawn if the system was not subjected to large changes, but they are not as strong as those that could be drawn if the same changes were made as part of a well designed experiment.

Although the broad distinction made between observational and experimental studies is useful, a little thought will show that both of these categories can be further subdivided in meaningful ways. For example, Eberhardt and Thomas (1991) propose a classification of studies into eight different types. However, this elaboration is unnecessary for the present discussion, where it merely needs to be noted that most environmental studies are observational, with all the potential limitations that this implies.

4.3 True Experiments and Quasi-Experiments

At this stage it is important to better define what is required for a study to be a 'true' experiment. Basically, the three important ingredients are randomization, replication, and controls.

Randomization should be used whenever there is an arbitrary choice to be made of which units will be measured out of a larger collection of possible units, or of the units to which different levels of a factor will be assigned. This does not mean that all selections of units and all allocations of factor levels have to be made completely at random. In fact, a large part of the theory of experimental design is concerned with how to restrict randomization and allocation in order to obtain the maximum amount of information from a fixed number of observations. Thus randomization is only required subject to whatever constraints are involved in the experimental design.

Randomization is used in the hope that it will remove any systematic effects of uncontrolled factors of which the experimenter has no knowledge. The effects of these factors will still be in the observations. However, randomization makes these effects part of the experimental errors that are allowed for by statistical theory. Perhaps more to the point, if randomization is not carried out then there is always the possibility of some unseen bias in what seems to be a haphazardous selection or allocation.

Randomization in experiments is not universally recommended. Its critics point to the possibility of obtaining random choices that appear to be unsatisfactory. For example, if different varieties of a crop are to be planted in different plots in a field then a random allocation can result in all of one variety being placed on one side of the field. Any fertility trends in the soil may then appear as a difference between the two varieties, and the randomization has failed to remove potential biases due to positions in the field. Although this is true, common sense suggests that if an experimenter has designed an experiment that takes into account all the obvious sources of variation, such as fertility trends in a field, so that the only choices left are between units that appear to be essentially the same, such as two plots in the same part of a field, then randomization is always worthwhile as one extra safeguard against the effects of unknown factors.

Replication is needed in order to decide how large effects have to be before they become difficult to account for in terms of normal variation. This requires the measurement of normal variation, which can be done by repeating experimental arrangements independently a number of times under conditions that are as similar as possible. Experiments without replication are case studies that may be quite informative and convincing, but it becomes a matter of judgement as to whether the outcome could have occurred in the absence of any manipulation.

Controls provide observations under normal conditions without the manipulation of factor levels. They are included in an experiment to give the standard with which the results under other conditions are compared. In the absence of controls it is usually necessary to assume that if there had been controls then these would have given a particular outcome. For instance, suppose that the yield of a new type of wheat is determined without running control trials with the standard variety. Then in order to decide whether the yield of the new variety is higher than that for the standard it is necessary to make some assumption about what the yield of the standard variety would have been under the experimental conditions. The danger here is obvious: it may be that under the conditions of the study the yield of the standard variety would not be what is expected, so that the new variety is being compared with the wrong standard.

Experiments that lack one or more of the ingredients of randomization, replication, and control are sometimes called quasi-experiments. Social scientists realized many years ago that they can often only do quasi-experiments rather that true experiments, and have considered very thoroughly the implications of this in terms of drawing conclusions (Campbell and Stanley, 1963; Manly, 1992, Chapter 5 and 6). An important point to realize is that many experiments in environmental science are really quasi-experiments, so that the social scientist's problems are also shared by environmental scientists.

There is no space here to discuss these problems at length. In fact, many of them are fairly obvious with a little thought. Some of the simpler designs that are sometimes used without a proper recognition of potential problems are listed below. Here O_i indicates observations made on a group of experimental units, X denotes an experimental manipulation, and R indicates a random allocation of experimental units to treatment groups. The description 'pre-experimental design' is used for the weakest situations, where inferences are only possible by making strong assumptions. 'Quasi-experimental designs' are better, but still not very satisfactory, while 'proper designs' have all the desirable characteristics of true experiments.

Two Pre-Experimental Designs

The one-group pretest-posttest design

$O_1 \quad X \quad O_2$

The two-group comparison without randomization

$X \quad O_1$

$\quad\;\; O_2$

A Quasi-Experimental Design

The comparative change design without randomization

$O_1 \quad X \quad O_2$

$O_3 \qquad\;\; O_4$

Two Proper Designs

The two-group comparison with randomization

$R \quad X \quad O_1$

$R \qquad\;\; O_2$

The comparative change design with randomization

$R \quad O_1 \quad X \quad O_2$

$R \quad O_3 \qquad\;\; O_4$

Of these designs, the comparative change ones are of particular interest because they are the same as the before-after-control-impact (BACI) design that is commonly used by environmental scientists, as in Example 1.4. Problems arise in these applications because it is not possible to randomly allocate experimental units to the treated and control groups before the treatment is applied.

4.4 Design-Based and Model-Based Inference

It is often not realized that conclusions from data are commonly reached using one of two very different philosophies. One is design-based, using the randomization used when collecting data, and the other is model-based, using the randomness that is inherent in the assumed model.

All of the methods for environmental sampling that were covered in Chapter 2 are design-based, because this is how the classical theory for sampling finite populations developed. For example, equation (3.4) for the variance of the mean of a random sample of size n from a population of size N with variance σ^2 states that

$$\mathrm{Var}(\bar{y}) = (\sigma^2/n)(1 - n/N).$$

What this means is that if the process of drawing a random sample is repeated many times, and the sample means $\bar{y}_1$, $\bar{y}_2$, $\bar{y}_3$, ... are recorded, then the variance of these means will be $(\sigma^2/n)(1 - n/N)$. Thus this is the variance that is generated by the sampling process. No model is needed for the distribution of the Y values, and in fact the variance applies for any distribution at all.

By way of contrast, consider the testing of the coefficient of X for a simple linear regression equation. In that case the usual situation is that there are n values $y_1, y_2, ..., y_n$ for Y with corresponding values x_1, $x_2, ..., x_n$ for X. The specific model

$$y_i = \beta_0 + \beta_1 x_i + \epsilon_i \quad (4.1)$$

is then assumed, where β_0 and β_1 are constants to be estimated, and ϵ_i is a random value from a normal distribution with a mean of zero and a constant unknown variance σ^2. The values of β_0 and β_1 are then estimated by least-squares as discussed in Section 4.4. If b_1 is the estimate of β_1, with an estimated standard error $\hat{SE}(b_1)$, then a test to see whether this is significantly different from zero involves comparing $b_1/\hat{SE}(b_1)$ with critical values of the t-distribution with n - 2 degrees of freedom.

In this case, there is no requirement that the units on which X and Y are measured are a random sample from some specific population. In fact, the equation for estimating the standard error of b_1 is based on the assumption that the X values for the data are fixed constants rather than being random, with the difference between b_1 and β_1 being due only to the particular values that are obtained for the errors ϵ_1, ϵ_2, ... ϵ_n in the model. Thus in this case the assessment about whether b_1 is significantly different from zero is not based on any requirement for random sampling of a population. Instead, it is based on the assumption that the model (4.1) correctly describes the structure of the data, and that the errors ϵ_1, ϵ_2, ... ϵ_n for the real data are a random sample from a normal distribution with mean zero and constant variance.

An advantage of the design-based approach is that valid inferences are possible which are completely justified by the design of the study and the way that data are collected. The conclusions can then always be defended providing that there is agreement about which variables should be measured, the procedures used to do the measuring, and the design protocol. In this case, any re-analysis of the data by other groups will not be able to declare these original conclusions incorrect. It is possible that a re-analysis using a model-based method may lead to different conclusions, but the original analysis will still retain its validity.

On the other hand, most statistical analysis is model-based, and there can be no question about the fact that this is necessary. The use of models allows much more flexibility in analyses and all of the methods described in the previous chapter are model-based, requiring specific assumptions about the structure of data. The flexibility comes at a price. Sometimes the implicit assumptions of models are hidden below the surface of the analysis. These range from assumptions about the random components of models that may or may not be critical as far as conclusions to assumptions about the mathematical form of the equations that relate different variables which may be absolutely critical particularly if it is necessary to predict the response of some variable when predictor variables are outside of the range observed on the available data. Moreover, whenever conclusions are drawn from a model-based analysis there is always the possibility that someone else will repeat the analysis with another equally reasonable model and reach different conclusions.

A case in point is the use of the lognormal distribution as a model for data. This is frequently assumed for the distribution of the concentration of a chemical in field samples because it has an appropriate shape (Figure 3.4). However, two recent studies have cast doubt on the uncritical use of this model.

Schmoyer *et al.* (1996) simulated data from lognormal, truncated normal and gamma distributions and compared the results from estimation and testing assuming a lognormal distribution with other approaches that do not make this assumption. They found that departures from the lognormal distribution were difficult to detect with the sample sizes that they used, but when they occurred the tests based on the lognormal assumption did not work as well as the alternatives. They concluded that "in the estimation of or tests about a mean, if the assumption of lognormality is at all suspect, then lognormal-based approaches may not be as good as the alternative methods". Because the lognormal distribution will probably seldom hold exactly for real data, this is a serious criticism of the model.

Wiens (1999) example is perhaps more disturbing. The same set of data was analysed two ways. First, the observations (the amount of antibody to hepatitis A virus in serum samples) were analysed assuming a generalized linear model with a possible mean difference between two groups, and lognormal errors. The difference in the two groups was approaching significance ($p = 0.10$) with the second group estimated to have a higher mean. Next, the data were analysed assuming gamma distributed errors, where the gamma distribution is one that often has the same type of shape as the lognormal. In this case the difference between the two groups was nowhere near significant ($p = 0.71$), but the first group was estimated to have a higher mean. Hence, the modelling assumptions made when analysing the data are rather crucial. Wiens notes that with this particular example a non-parametric test can be used to compare the two groups, which is better than the model-based approach. However, he also points out that with more complicated data sets a model-based approach may be preferred because of the flexibility that this permits in the analysis. He therefore proposes the ad-hoc solution of analysing data like this using both the lognormal and gamma models and investigating further if the results do not agree. This, of course, raises the possibility that both models are wrong, with similar misleading outcomes that go undetected.

The moral from all this is that although a strict adherence to design-based analyses is not possible for environmental studies, it is a good idea to rely on design-based analyses as much as possible. The value of at least a few indisputable design-based statistical inferences may, for example, be of great value for defending a study in a court case.

4.5 Tests of Significance and Confidence Intervals

The concept of a test of significance is discussed in Section A4 of Appendix A. Such tests are very commonly used for drawing conclusions from data. However, they do have certain limitations which have led over the years to a number of authors questioning their use, or at least the extent to which they are used.

The two basic problems were mentioned in Example 1.7 which is concerned with the comparison of the mean of a variable at a site which was once contaminated and is now supposed to be cleaned up with the mean at a reference site which was never contaminated. The first problem is that the two sites cannot be expected to have exactly the same mean even if the cleaning operation has been very effective. Therefore, if large samples are taken from each site there will be a high probability of finding a significant difference between the two sample means. The second problem is that if the difference between the sample means is not significant, then it does not mean that no difference exists. An alternative explanation is that the sample sizes used are not large enough to detect the existing differences.

These well-known problems have been discussed many times by social scientists (e.g., Oakes, 1986), medical statisticians (e.g., Gardner and Altman, 1986), environmental scientists (e.g., McBride *et al.*, 1993), wildlife scientists (e.g., Cherry, 1998 and Johnson, 1999), statisticians in general (e.g., Nelder, 1999), and no doubt by those working in other areas as well. A common theme is that too often hypothesis tests are used when it is obvious in advance that the null hypothesis is not true, and that as a result scientific papers are becoming cluttered up with unnecessary p-values.

In truth, there is not much point in testing hypotheses that are known to be false. Under such circumstances it makes more sense to estimate the magnitude of the effect of interest, with some indication of the likely accuracy of results. However, there are situations where the truth of a null hypothesis really is in question, and then carrying out a significance test is an entirely reasonable thing to do. Once evidence for the existence of an effect is found, it is then reasonable to start measuring its magnitude.

Testing null models in ecology is a case in point, as discussed in the book by Gotelli and Graves (1996). The point of view adopted is that there is a whole range of situations where it is at least plausible that ecological structures are just the result of chance. For example, if a number of islands in the same general area are considered, then the species that are present on each island can be recorded. A null model then says that the species occur on different islands completely independent of each other, possibly subject to certain constraints such as the sizes of the islands and the fact that some species are naturally more widespread than others. This null model can then be tested by choosing a test statistic S which measures in some sense the apparent interaction between species occurrences, and seeing whether the observed value of S could reasonably have occurred if the null hypothesis is true.

Some ecologists do not seem to like the idea of testing null models like this, perhaps because they believe that it is obvious that species interact. However, proponents of these methods do not accept that this is necessarily true and argue that testing this fundamental idea is still necessary. From their point of view tests of significance are an entirely reasonable tool for analysing data.

No doubt arguments about the value of tests of significance will continue. The point of view adopted here is that it does often happen that the existence of an effect is in doubt, in which case testing the null hypothesis that the effect does not exist is sensible. However, in other cases it is more or less certain that an effect exists and the main question of interest is the size of the effect. In that case a confidence interval may provide the necessary information. Thus both tests of significance and confidence intervals are important tools for data analysis, but under different circumstances.

4.6 Randomization Tests

Randomization inference is a computer-intensive method that is receiving more use as time goes by for the analysis of environmental data. It has a long history, going back about 65 years to the work of Sir Ronald Fisher, one of developers of many of the statistical methods used today (Fisher, 1935, 1936). The method is referred to quite often in the following chapters. It was also used in the analyses of data from the Exxon Shoreline Ecology Program and the Biological Monitoring Survey that are discussed in Example 1.1, and in Carpenter *et al.*'s (1989) analyses of large-scale perturbation experiments that are discussed in Example 1.4.

The simplest situation for understanding what is meant by a randomization test is the two-group comparison, as proposed by Fisher (1936). In this situation there is one sample of values $x_1, x_2, ..., x_m$, with mean $\bar{x}$, and a second sample of values $y_1, y_2, ..., y_n$, with mean $\bar{y}$. The question of interest is whether the two samples come from the same distribution or, more precisely, whether the absolute mean difference $|\bar{x} - \bar{y}|$ is small enough for this to be plausible.

The test proceeds as follows:

(1) The observed absolute mean difference is labelled d_1.

(2) It is argued that if the null hypothesis is true (the two samples come from the same distribution), then any one of the observed values $x_1, x_2, ..., x_m$ and $y_1, y_2, ..., y_n$ could equally well have occurred in either of the samples. On this basis, a new sample 1 is chosen by randomly selecting m out of the full set of n + m values, with the remaining values providing the new sample 2. The absolute mean difference $d_2 = |\bar{x} - \bar{y}|$ is then calculated for this randomized set of data.

(3) Step (2) is repeated a large number R - 1 of times to give a total of R differences $d_1, d_2, ..., d_R$.

(4) The R differences are put in order from the smallest to largest.

(5) If the null hypothesis is true, then d_1 should look like a typical value from the set of R differences, and is equally likely to appear anywhere in the list. On the other hand, if the two original samples come from distributions with different means, then d_1 will tend to be near the top of the list. On this basis, d_1 is said to be significantly large at the $100\alpha\%$ level if it is among the top $100\alpha\%$ of values in

the list. If 100α% is small (say 5% or less), then this is regarded as evidence against the null hypothesis.

It is an interesting fact that this test is exact in a certain sense even when R is quite small. For example, suppose that R = 99. Then if the null hypothesis is true and there are no tied values in the differences d_1, d_2, ..., d_{100}, the probability of d_1 being one of the largest 5% of values (i.e., one of the largest 5) is exactly 0.05. This is precisely what is required for a test at the 5% level: the probability of a significant result when the null hypothesis is true is equal to 0.05.

The test just described is two-sided. A one-sided version is easily constructed by using the signed difference $\bar{x} - \bar{y}$ as the test statistic, and seeing whether this is significantly high (assuming that the alternative to the null hypothesis of interest is that the values in the first sample come from a distribution with a higher mean than that for the second sample).

An advantage that the randomization approach has over a conventional parametric test on the sample mean difference is that it is not necessary to assume any particular type of distribution for the data, such as normal distributions for the two samples for a t-test. The randomization approach also has an advantage over a non-parametric test like the Mann-Whitney U-test because it allows the original data to be used rather than just the ranks of the data. Indeed, the Mann-Whitney U-test is really just a type of randomization test for which the test statistic only depends on the ordering of the data values in the two samples being compared.

A great deal more could be said about randomization testing, and much fuller accounts are given in the books by Edgington (1987), Good (1994) and Manly (1997a). The following example shows the outcome obtained on a real set of data.

Example 4.1 Survival of Rainbow Trout

This example concerns part of the results shown in Table 3.8 from a series of experiments conducted by Marr *et al.* (1995) to compare the survival of naive and metals-acclimated juvenile brown trout (*Salmo trutta*) and rainbow trout (*Oncorhynchus mykiss*) when exposed to a metals mixture with the maximum concentrations found in the Clark Fork River, Montana, USA. Here only the results for the hatchery rainbow trout will be considered. For these there were 30 control fish that were randomly selected to be controls that were kept in clean water for three weeks before being transferred to the metals mixture, while the remaining 29 fish were acclimated for three weeks in a weak solution of metals before being transferred to the stronger mixture. All the fish survived the initial three week period, and there is interest in whether the survival time of the fish in the stronger mixture was affected by the treatment.

In Example 3.2 the full set of data shown in Figure 4.8 was analysed using two factor analysis of variance. However, a logarithmic transformation was applied to the survival times first to largely overcome the tendency for the standard deviation of survival times to increase with the mean. This is not that apparent when only the hatchery rainbow trout are considered. Therefore if only these results were known, then it is quite likely that no transformation would be considered necessary. Actually, as will be seen shortly it is immaterial whether a transformation is made or not if only these fish are considered.

Note that for this example, carrying out a test of significance to compare the two samples is reasonable. Before the experiment was carried out It was quite conceivable that the acclimation would have very little, if any, effect on survival, and it is interesting to know whether the observed mean difference could have occurred purely by chance.

The mean survival difference (acclimated - control) is 10.904. Testing this using the randomization procedure using steps (1) to (5) described above using 4999 randomizations resulted in the observed value of 10.904 being the largest in the set of 5000 absolute mean differences consisting of itself and the 4999 randomized differences. The result is therefore significantly different from zero at the 0.02% level (1/5000). Taking logarithms to base 10 of the data and then running the test gives a mean difference of 0.138, which is again significantly different from zero at the 0.02% level. There is very strong evidence that acclimation affects the mean survival time, in the direction of increasing it.

Note that if it was decided in advance that if anything acclimation would increase the mean survival time then the randomization test could be made one-sided to see whether the acclimated - control mean difference is significantly large. This also gives a result that is significant at the 0.2% level either using the survival times or logarithms of these.

Note also that in this example the use of a randomization test is completely justified by the fact that the fish were randomly allocated to a control and acclimated group before the experiment began. This ensures that if acclimation has no effect then the data that are observed are exactly equivalent to one of the alternative sets generated by randomizing this observed data. Any one of the randomized sets of data really would have been just as likely to occur as the observed set. When using a randomization test it is always desirable that an experimental randomization is done to fully justify the test, although this is not always possible.

Finally, it is worth pointing out that any test on these data can be expected to give a highly significant result. A two-sample t-test using the mean survival times and assuming that the samples are from populations with the same variance gives $t = 5.70$ with 57 degrees of freedom, giving $p = 0.0000$. Using logarithms instead gives $t = 5.39$ still with 57 degrees of freedom and $p = 0.0000$. Using a Mann-Whitney U-test either on the survival times or logarithms also gives $p = 0.0000$. The conclusion is therefore the same, whatever reasonable test is used.

4.7 Bootstrapping

Bootstrapping as a general tool for analysing data was first proposed by Efron (1979). Initially the main interest was in using this method to construct confidence intervals for population parameters using the minimum of assumptions, but more recently there has been interest in bootstrap tests of hypotheses (Efron and Tibshirani, 1993; Hall and Wilson, 1991; Manly, 1997a).

The basic idea behind bootstrapping is that when only sample data are available, and no assumptions can be made about the distribution that the data are from, then the best guide to what might happen by taking more samples from the distribution is provided by resampling the sample. This is a very general idea, and the way that it might be applied is illustrated by the following example.

Example 4.2 A Bootstrap 95% Confidence Interval

Table 3.3 includes the values for chlorophyll-a for 25 lakes in a region. Suppose that the total number of lakes in the region is very large and that there is interest in calculating a 95% confidence interval for the mean of chlorophyll-a, assuming that the 25 lakes are a random sample of all lakes.

If the chlorophyll-a values were approximately normally distributed, then this would probably be done using the t-distribution using equation (A9) from Appendix A. The interval would then be

$$\bar{x} - 2.064s/\sqrt{25} < \mu < \bar{x} + 2.064s/\sqrt{25}, \qquad (4.2)$$

where $\bar{x}$ is the sample mean, s is the sample standard deviation, and 2.064 is the value that is exceeded with probability 0.025 for the t-distribution with 24 degrees of freedom. For the data in question, $\bar{x}$ = 50.30 and s = 50.02, so the interval is

$$29.66 < \mu < 70.95.$$

The problem with this is that the values of chrorophyll-a are very far from being normally distributed, as is clear from Figure 4.1. There is therefore a question about whether this method for determining the interval really gives the required level of 95% confidence.

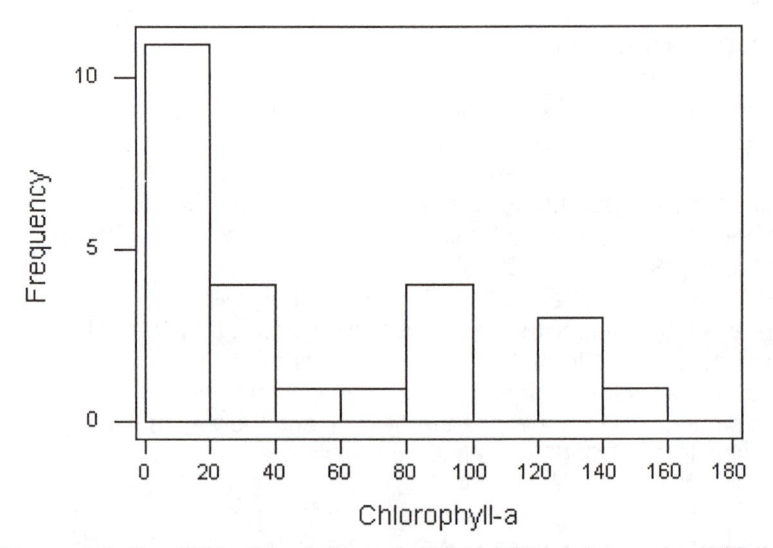

Figure 4.1 The distribution of chlorophyll-a for 25 lakes in a region, with the height of the histogram bars reflecting the percentage of the distribution in different ranges.

Bootstrapping offers a possible method for obtaining an improved confidence interval, with the method that will now be described being called bootstrap-t (Efron, 1981). This works by using the bootstrap to approximate the distribution of

$$t = (\bar{x} - \mu)/(s/\sqrt{25})$$

instead of assuming that this follows a t-distribution with 24 degrees of freedom, which it would for a sample from a normal distribution. An algorithm to do this is as follows, where this was easily carried out in a spreadsheet program:

(1) The 25 sample observations of chlorophyll-a from Table 4.3 are set up as the bootstrap population to be sampled. This population has the known mean of $\mu_B = 50.30$.

(2) A bootstrap sample of size 25 is selected from the population by making each value in the sample equally likely to be any of the 25 population values. This is sampling with replacement, so that a population value may occur 0, 1, 2, 3 or more times.

(3) The t-statistic $t_1 = (\bar{x} - \mu_B)/(s/\surd 25)$ is calculated from the bootstrap sample.

(4) Steps (2) and (3) are repeated 5000 times to produce 5000 t-values $t_1, t_2, ..., t_{5000}$ to approximate the distribution of the t-statistic for samples from bootstrap population.

(5) Using the bootstrap distribution obtained, two critical values t_{low} and t_{high} are estimated such that

$$\text{Prob}[(\bar{x} - \mu_B)/(s/\surd 25) < t_{low}] = 0.025,$$

and

$$\text{Prob}[(\bar{x} - \mu_B)/(s/\surd 25) > t_{high}] = 0.025.$$

(6) It is assumed that the critical values also apply for random samples of size 25 from the distribution of chlorophyll-a from which the original set of data was drawn. Thus it is asserted that

$$\text{Prob}[t_{low} < (\bar{x} - \mu)/(s/\surd 25) < t_{high}] = 0.95,$$

where $\bar{x}$ and s are now the values calculated from the original sample, and μ is the mean chlorophyll-a value for all lakes in the region of interest. Rearranging the inequalities then leads to the statement that

$$\text{Prob}[\bar{x} - t_{high}s/\surd 25 < \mu < \bar{x} - t_{low}s/\surd 25] = 0.95,$$

so that the required 95% confidence interval is

$$\bar{x} - t_{high}s/\surd 25 < \mu < \bar{x} - t_{low}s/\surd 25. \quad (4.3)$$

The interval (4.3) only differs to the extent that t_{low} and t_{high} vary from 2.064. When the process was carried out it was found that the bootstrap distribution of $t = (\bar{x} - \mu_B)/(s/\sqrt{25})$ is quite close to the t-distribution with 24 degrees of freedom, as shown by Figure 4.2, but with $t_{low} = -2.6$ and $t_{high} = 2.0$. Using the sample mean and standard deviation, the bootstrap-t interval therefore becomes

$$50.30 - 2.0(50.02/5) < \mu < 50.30 + 2.6(50.02/5),$$

or

$$30.24 < \mu < 76.51.$$

These compare with the limits of 29.66 to 70.95 obtained using the t-distribution. Thus the bootstrap-t method gives a rather higher upper limit, presumably because this takes better account of the type of distribution being sampled.

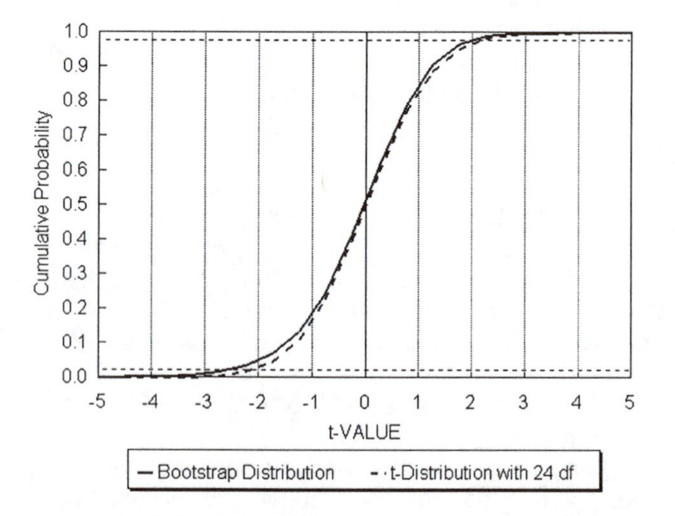

Figure 4.2 Comparison between the bootstrap distribution of $(\bar{x} - \mu_B)/(s/\sqrt{25})$ and the t-distribution with 24 degrees of freedom. According to the bootstrap distribution, the probability of a value less than $t_{low} = -2.6$ is approximately 0.025, and the probability of a value higher than $t_{high} = 2.0$ is approximately 0.025.

4.8 Pseudoreplication

The term "pseudoreplication" causes some concern, particularly among field ecologists, with the clear implication that when an investigator believes that replicated observations have been taken, this may not really the case at all. Consequently, there is some fear that the conclusions from studies will not be valid because of unrecognized pseudoreplication.

The concept of pseudoreplication was introduced by Hurlbert (1984) with the definition "the use of inferential statistics to test for treatment effects with data from experiments where either treatments are not replicated, or replicates are not statistically independent". Two examples of pseudoreplication are:

- A sample of metre square quadrats randomly located within a 1 ha study region randomly located in a larger burned area is treated as a random sample from the entire burned area.

- Repeated observations on the location of a radio-tagged animal are treated as a simple random sample of the habitat points used by the animal, although in fact successive observations tend to be close together in space.

In both of these examples it is the application of inferential statistics to dependent replicates as if they were true replicates from the population of interest that causes the pseudoreplication. However, it is important to understand that using a single observation per treatment or replicates that are not independent data is not necessarily wrong. Indeed it may be unavoidable in some field studies. What is wrong is to ignore this in the analysis of the data.

There are two common aspects of pseudoreplication. One of these is the extension of a statistical inference observational study beyond the specific population studied to other unstudied populations. This is the problem with the first example above on sampling of burned areas. The other aspect is the analysis of dependent data as if they are independent data. This is the problem with the example on radio-tagged animals.

When dependent data are analysed as if they are independent, the sample size used is larger than the effective number of independent observations. This often results in too many significant results being obtained from tests of significance, and confidence intervals being narrower than they should be. To avoid this, a good rule to follow is that statistical inferences should be based on only one value from each independently sampled unit, unless the dependence in the data is properly handled in the analysis. For example, if five quadrats are randomly located in a study area, then statistical inferences about the area should be based on five values, regardless of the number of plants, animals, soil samples, etc., that are counted or measured in each quadrat. Similarly, if a study uses data from 5 radio-tagged animals, then statistical inferences about the population of animals should be based on a sample of size 5, regardless of the number of times each animal is relocated.

When data are dependent because they are collected close together in time or space there are a very large number of analyses available to allow for this. Many of these methods are discussed in later chapters in connection with particular types of applications, particularly in Chapters 8 and 9. For now it is just noted that unless it is clearly possible to identify independent observations from the study design, then one of these methods needs to be used.

4.9 Multiple Testing

Suppose that an experimenter is planning to run a number of trials to determine whether a chemical at a very low concentration in the environment has adverse effects. A number of variables will be measured (survival times of fish, growth rate of plants, etc.) with comparisons between control and treated situations, and the experimenter will end up doing 20 tests of significance, each at the 5% level. He decides that if any of these tests give a significant result then there is evidence of adverse effects. This experimenter has a multiple testing problem.

To see this, suppose that the chemical has no perceptible effects at the level tested so that the probability of a significant effect on any one of his 20 tests is 0.05. Suppose also that the tests are on independent data. Then the probability of none of the tests being significant is $0.95^{20} = 0.36$, so that the probability of obtaining at least one significant result is $1 - 0.36 = 0.64$. Hence the likely outcome of the experimenter's work is to conclude that the chemical has an adverse effect even when it is harmless.

Many solutions to the multiple testing problem have been proposed. The best known of these relate to the specific problem of comparing the mean values at different levels of a factor in conjunction with analysis of variance. These are discussed at length in general statistics texts (e.g., Steel and Torrie, 1980, Chapter 8; Underwood, 1997, Section 8.6), and also in the more specialised texts of Hochberg and Tamhane (1987) and Westfall and Young (1993). These multiple comparison procedures are also available in standard statistical computer packages, although they are not accepted as necessarily being useful by all statisticians (e.g., Saville, 1986, 1990; Mead, 1988, p. 311; Nelder, 1999).

There are also some procedures that can be applied more generally when several tests are to be conducted at the same time. Of these, the Bonferroni procedure is the simplest. This is based on the fact that if m tests are carried out at the same time using the significance level $(100\alpha\%)/m$, and all of the null hypotheses are true, then the probability of getting any significant result is less than α. Thus the experimenter with 20 tests to carry out can use the significance level (5%)/20 = 0.25% for each test, and this ensures that the probability of getting any significant results is less than 0.05 when no effects exist.

An argument against using the Bonferroni procedure is that it requires very conservative significance levels when there are many tests to carry out. This has led to the development of a number of improvements that are designed to result in more power to detect effects when they do really exist. Of these, the method of Holm's (1979) appears to be the one which is easiest to apply (Peres-Neto, 1999). This does not however take into account the correlation between the results of different tests. If some correlation does exist because the different test statistics are based partially on the same data, then, in principle, methods which allow for this should be better, such as the randomization procedure described by Manly (1997a, Section 6.8) which can be applied in a wide variety of different situations (e.g., Holyoak and Crowley, 1993), or several approaches that are described by Troendle and Legler (1998).

Holm's method works using the following algorithm:

(a) Decide on the overall level of significance α to be used (the probability of declaring anything significant when the null hypotheses are all true).

(b) Calculate the p-value for the m tests being carried out.

(c) Sort the p-values into the ascending order, to give p_1, p_2, ..., p_m, with any tied values being put in a random order.

(d) See if $p_1 \leq \alpha/m$, and if so declare the corresponding test to give a significant result, otherwise stop. Next see if $p_2 \leq \alpha/(m - 1)$, and if so declare the corresponding test to give a significant result, otherwise stop. Next see if $p_3 \leq \alpha/(m - 2)$, and if so declare the corresponding test to give a significant result, otherwise stop. Continue this process until an insignificant result is obtained, or until it is seen whether $p_k \leq \alpha$, in which case the corresponding test is declared to give a significant result. Once an insignificant result is obtained, all the remaining tests are also insignificant, because their p-values are at least as large as the insignificant one.

The procedure is illustrated by the following example.

Example 4.3 Multiple Tests on Characters for Brazilian Fish

This example to illustrate the Holm (1979) procedure is the one also used by Peres-Neto (1999). The situation is that five morphological characters have been measured for 47 species of Brazilian fish and there is interest in which pairs of characters show significant correlation. Table 4.1 shows the ten pairwise correlations obtained with their probability values based on the assumptions that the 47 species of fish are a random sample from some population, and that the characters being measured have normal distributions for this population. (For the purposes of this example the validity of these assumptions will not be questioned.)

Table 4.1 Correlations (r) between characters for 47 species of Brazilian fish, with corresponding p-values (e.g., the correlation between characters 1 and 2 is 0.110, with p = 0.460)

		Character			
Character		1	2	3	4
2	r	0.110			
	p-value	0.460			
3	r	0.325	0.345		
	p-value	0.026	0.018		
4	r	0.266	0.130	0.142	
	p-value	0.070	0.385	0.340	
5	r	0.446	0.192	0.294	0.439
	p-value	0.002	0.196	0.045	0.002

The calculations for Holm's procedure, using an overall significance level of 5% ($\alpha = 0.05$) are shown in Table 4.2. It is found that just two of the correlations are significant after allowing for multiple testing.

Table 4.2 Calculations and results from the Holm (1979) method for multiple testing using the correlations and p-values from Table 4.1, and $\alpha = 0.05$

i	r	p-value	0.05/(m+i-1)	Significance
1	0.439	0.002	0.005	yes
2	0.446	0.002	0.006	yes
3	0.345	0.018	0.006	no
4	0.325	0.026	0.007	no
5	0.294	0.045	0.008	no
6	0.266	0.070	0.010	no
7	0.192	0.196	0.013	no
8	0.142	0.340	0.017	no
9	0.130	0.385	0.025	no
10	0.110	0.460	0.050	no

4.10 Meta-Analysis

The term 'meta-analysis' is used to describe methods for combining the results from several studies to reach an overall conclusion. This can be done in a number of different ways, with the emphasis either on determining whether there is overall evidence of the effect of some factor, or of producing the best estimate of an overall effect.

A simple approach to combining the results of several tests of significance was proposed by Fisher (1970). This is based on three well-known results:

(a) if the null hypothesis is true for a test of significance, then the p-value from the test has a uniform distribution between 0 and 1 (i.e., any value in this range is equally likely to occur);

(b) if p has a uniform distribution, then $-2\log_e(p)$ has a chi-squared distribution with 2 degrees of freedom; and

(c) if $X_1, X_2, ..., X_n$ all have independent chi-squared distributions then their sum, $S = \sum X_i$ also has a chi-squared distribution, with the number of degrees of freedom being the sum of the degrees of freedom for the components.

It follows from these results that if n tests are carried out on the same null hypothesis using independent data and yield p-values of $p_1, p_2, ..., p_n$, then a sensible way to combine the test results involves calculating

$$S_1 = -2 \sum \log_e(p_i), \tag{4.4}$$

where this will have a chi-squared distribution with 2n degrees of freedom if the null hypothesis is true for all of the tests. A significantly large value of S_1 is evidence that the null hypothesis is not true for at least one of the tests. This will occur if one or more of the individual p-values is very small, or if most of the p-values are fairly small.

There are a number of alternative methods that have been proposed for combining p-values, but Fisher's method seems generally to be about the best providing that the interest is in whether the null hypothesis is false for any of the sets of data being compared (Folks, 1984). However, Rice (1990) argued that sometimes this is not quite what is needed. Instead, the question is whether a set of tests of a null hypothesis are in good agreement about whether there is evidence against the null hypothesis. Then a consensus p-value is needed to indicate whether, on balance, the null hypothesis is supported or not. For this purpose Rice suggests using the Stouffer method described by Folks (1984).

The Stouffer method proceeds as follows. First, the p-value from each test is converted to an equivalent z-score, i.e., the p-value p_i for the ith test is used to find the value z_i such that

$$\text{Prob}(Z < z_i) = p_i, \tag{4.5}$$

where Z is a random value from the standard normal distribution with a mean of zero and a standard deviation of one. If the null hypothesis is true for all of the tests, then all of the z_i values will be random values from the standard normal distribution, and it can be shown that their mean $\bar{z}$ will be normally distributed with a mean of zero and a variance of $1/\sqrt{n}$. The mean z-value can therefore be tested for significance by seeing whether

$$S_2 = \bar{z} / (1/\sqrt{n}) \tag{4.6}$$

is significantly less than zero.

There is a variation on the Stouffer method that is appropriate when for some reason it is desirable to weight the results from different studies differently. This weighting might, for example, be based on the sample sizes used in the study, some assessment of reliability, or perhaps with recent studies given the highest weights. This is called the Liptak-Stouffer method by Folks 1984). In this case, let w_i be the weight for the ith study, and define the test statistic

$$S_3 = (w_1 z_1 + w_2 z_2 + \dots + w_n z_n)/\sqrt{(w_1^2 + w_2^2 + \dots + w_n^2)}. \tag{4.7}$$

If the null hypothesis is true for all studies, then this will follow a standard normal distribution. If it is significantly low in comparison with the standard normal distribution, then this is evidence that the null hypothesis is not always true.

Meta-analysis as generally understood involves more than just combining the p-values from several sets of data. In fact, the usual approach is to take a series of studies and for each one calculate an estimated effect size, which is often just the mean difference between the treated and control groups in units of the estimated standard deviation of individual observations. Questions of interest are:

- How large is the effect overall?
- Is it generally a positive or a negative effect, and is it usually different from zero?
- Are the effect sizes similar for all studies?
- If there is variation between studies, can this be related to the different types of studies involved?

There is a large literature on this type of meta-analysis. Comprehensive sources for more information are the books by Hedges and Olkin (1985, 1999), while the introductions to the topic provided by Gurevitch and Hedges (1993, 1999) will be useful for beginners in this area. Also, a recent Special Feature in the journal *Ecology* gives an up-to-date review of applications of meta-analysis in this area (Osenberg *et al.*, 1999).

Example 4.4 The **Exxon Valdez** *Oil Spill and Intertidal Sites*

Table 4.3 summarises the results from part of the Oil Spill Trustees' Coastal Habitat Injury Assessment Study of the shoreline impact of the *Exxon Valdez* oil spill that was discussed in Example 1.1. The source of these results is the paper by McDonald *et al.* (1995), who also provide the example calculations considered here. They come from comparing the biomass of barnacles (the family *Balanomorpha*) at some paired intertidal sites, of which one was oiled and the other not. The pairing was designed to ensure that the sites were similar in all respects except the oiling.

Table 4.3 Results of comparing the biomass of mussels at paired oiled and control sites on Prince William Sound, Alaska, in the second metre of vertical drop during May 1990

Site pair	Density (Count/0.1m^2) Unoiled	Oiled	p-value from test	Inclusion probability (q)
1	22.7	13.2	0.3740	1.000
2	62.8	32.0	0.0146	0.262
3	428.4	232.7	0.1670	0.600
4	385.9	132.7	0.0515	0.179
5	213.7	112.9	0.1479	0.668

One way of looking at these results is to regard each pair of sites as a quasi-experiment on the effects of oiling. Each experiment can then be summarised by its p-value, which gives the probability of observing a positive difference between the biomass for the control and oiled sites as large as that observed. This is the p-value for a one-sided test because the assumption is that, if anything, oiling will reduce the biomass. The p-values are 0.374, 0.015, 0.167, 0.052 and 0.148. They are all fairly small, but only one gives significance at the 5% level. A reasonable question is whether all the sites between them give real evidence of an effect of oiling.

This question can be answered using Fisher's (1970) method for combining p-values from independent data. From equation (4.4) the statistic for the combined p-values is

$$S_1 = -2\{\log_e(0.3740) + \log_e(0.0146) + \log_e(0.1673) + \log_e(0.0515) + \log_e(0.1479)\}$$

$$= 23.75,$$

with 2 x 5 = 10 degrees of freedom. The probability of a value this large from the chi-squared distribution with 10 degrees of freedom is 0.0083, which is the combined p-value. Consequently, taken together the five sites give a result that is significant at the 1% level (since $0.0083 < 0.01$) and there is clear evidence of an effect of oiling at one or more of the sites.

A problem with this conclusion is that it says nothing about intertidal sites in general. However, a conclusion in that respect is possible using the Liptak-Stouffer method. Consider the ith site, with the p-value p_i, and the corresponding z-score z_i from equation (4.5). The z-score can then be thought of as an observation for the site pair, which is one unit from a population consisting of all the site pairs that could have been sampled. If the sample of five site pairs used in the study were a random sample from the population of all site pairs then the mean z-score for the sample would be an unbiased estimator of the population mean, allowing an inference to be drawn immediately about oiling in general. In particular, if the sample mean is significantly less than zero, then this would be evidence of an oiling effect in general.

Unfortunately, the site pairs were not a random sample from the population of all possible site pairs because the pairs were made up after a stratified sampling plan was found to be unsatisfactory because a high number of sites were found to be in the wrong strata by field teams. However, this problem can be overcome by using the Horvitz-Thomson estimator described in Section 2.14, and equation (2.51) in particular. Thus

$$\hat{\mu}_z = (z_1/q_1 + z_2/q_2 + ... + z_5/q_5)/(1/q_1 + 1/q_2 + ... +1/q_5),$$

is an unbiased estimator of the population mean of z, where q_i is the probability of the ith site being included in the sample. Setting

$$w_i = (1/q_i)/(1/q_1 + 1/q_2 + ... +1/q_5),$$

makes

$$\hat{\mu}_z = w_1z_1 + w_2z_2 + ... + w_5z_5,$$

which will be normally distributed with a mean of zero and a variance of $w_1^2 + w_2^2 + ... +w_5^2$ if the null hypothesis of no oiling effects on average is true. Hence this null hypothesis can be tested by seeing whether

$$S_3 = (w_1z_1 + w_2z_2 + ... + w_5z_5)/\sqrt{(w_1^2 + w_2^2 + ... +w_5^2)}$$

is significantly less than zero in comparison with the standard normal distribution. This is precisely the Liptak-Stouffer test statistic of equation (4.7) showing that one application of the Liptak-Stouffer method is where the weight used for a z-value is proportional to the reciprocal of the probability of the observation being included in the sample that is available.

The inclusion probabilities for the five sites are shown in the last column of Table 4.3. Using these probabilities, the calculation of S_3 is shown in Table 4.4. It is found that $S_3 = -2.91$, where the probability of a value this low or lower for the standard normal distribution is

0.0018. This is therefore significantly low at the 1% level, giving strong evidence of an average effect of oiling.

Table 4.4 Calculation of the Liptak-Stouffer test statistic for the data in Table 4.3 on the biomass of mussels at oiled and unoiled sites

Site pair	Inclusion probability (q)	Weight (w)[1]	Test result (p)	Normal score (z)[2]	Product wz
1	1.000	0.0737	0.3740	-0.321	-0.0237
2	0.262	0.2813	0.0146	-2.181	-0.6135
3	0.600	0.1228	0.1673	-0.966	-0.1187
4	0.179	0.4118	0.0515	-1.630	-0.6714
5	0.668	0.1103	0.1479	-1.045	-0.1154
Sum		1.0000			-1.5426
$\sum w^2$		0.2814			
				Liptak-Souffer statistic[3]	-2.9080

[1]The weight for site i is $(1/q_i)/\sum(1/q_k)$.
[2]The normal score is the value obtained from equation (4.5).
[3]Calculated as $\sum w_i z_i/\sqrt{\{\sum w_i^2\}}$.

This analysis is still not without problems. In particular, the z values are only a very indirect measure of the difference between oiled and unoiled sites, and all that can be said if there is an effect of oiling is that these z-values will tend to be negative. An analysis based on a more meaningful measure of the effect of oiling would therefore have been preferable.

4.11 Bayesian Inference

So far the methods discussed in this book have all been based on a traditional view of statistics, with tests of significance and confidence intervals being the main tools for inference, justified either by the study design (for design-based inference) or an assumed model (with model-based inference). There is, however, another fundamentally different approach to inference that is being used increasingly in recent times because certain computational difficulties that used to occur have now been overcome.

This alternative approach is called Bayesian inference because it is based on a standard result in probability theory called Bayes' theorem. To see what this theorem says, consider the situation where it is possible to state that a certain parameter θ must take one, and only one, of a set of n specific values denoted by θ_1, θ_2, ... θ_n, and where before any data are collected it is known that the prior probability of θ_i (i.e. the probability of this being the correct value for θ) is $P(\theta_i)$, with $P(\theta_1) + P(\theta_2) + ... + P(\theta_n) = 1$. Some data that are related to θ are then collected. Under these conditions, Bayes' theorem states that the posterior probability that θ_i (i.e., the probability that this is the correct value of θ, given the evidence in the data) is

$$P(\theta_i|\text{data}) = P(\text{data}|\theta_i)/ \sum P(\text{data}|\theta_k)P(\theta_k), \tag{4.8}$$

where the summation is for k = 1 to n, and $P(\text{data}|\theta_i)$ is the probability of observing the data if $\theta = \theta_i$.

This result offers a way to calculate the probability that a particular value is the correct one for θ on the basis of the data, which is the best that can be hoped for in terms of inferences about θ. The stumbling block is that in order to do this it is necessary to know the prior probabilities before the data are collected.

There are two approaches used to determine prior probabilities when, as is usually the case, these are not really known. The first approach uses the investigator's subjective probabilities, based on general knowledge about the situation. The obvious disadvantage of this is that another investigator will likely not have the same subjective probabilities so that the conclusions from the data will depend to some extent at least on who does the analysis. It is also very important that the prior probabilities are not determined after the data have been examined because equation (4.8) does not apply if the prior probabilities depend on the data. Thus inferences based on equation (4.8) with the prior probabilities depending partly on the data are not Bayesian inferences. In fact, they are not justified at all.

The second approach is based on choosing prior probabilities that are uninformative, so that they do not have much effect on the posterior probabilities. For example, the n possible values of θ can all be given the prior probability 1/n. One argument for this approach is that it expresses initial ignorance about the parameter in a reasonable way, and that, providing there is enough data, the posterior probabilities do not depend very much on whatever is assumed for the prior probabilities.

Equation (4.8) generalizes in a straightforward way to situations where there are several or many parameters of interest, and where the prior distributions for these parameters are discrete or continuous. For many purposes, all that needs to be known is that

$$P(\text{parameters}|\text{data}) \propto P(\text{data}|\text{parameters})P(\text{parameters}),$$

i.e., the probability of a set of parameter values given the data is proportional to the probability of the data given the parameter values, multiplied by the prior probability of the set of parameter values. This result can be used to generate posterior probability distributions using possibly very complicated models when the calculations are done using a special technique called Markov chain Monte Carlo.

There is one particular aspect of Bayesian inference that should be appreciated. It is very much model-based in the sense discussed in Section 4.4. This means that it is desirable with any serious study that the conclusions from an analysis should be quite robust to both the assumptions made about prior distributions, and the assumptions made about the other components in the model. Unfortunately, these types of assessments are often either not done, or not done very thoroughly.

This brief introduction to Bayesian inference has been included here because it seems likely that environmental scientists will be seeing more of these methods in the future as means of drawing conclusions from data. More information about them with the emphasis on Markov Chain Monte Carlo methods is provided in the books by Manly (1997a) and Gilks *et al.* (1996). For more on Bayesian data analysis in general see the book by Gelman *et al.* (1995), and for arguments why these approaches should be viewed with caution see Dennis (1996).

4.12 Chapter Summary

- In drawing conclusions from data, an important distinction is between observational and experimental studies. In general, observational studies are more likely to be affected by uncontrolled factors, leading to incorrect conclusions.

- Experimental studies can be either true experiments, or quasi-experiments. True experiments incorporate randomization, replication and controls, while quasi-experiments lack one of these components. Many studies in environmental science are really quasi-experiments, so it is important to realize the limitations that this imposes on inferences.

- There are two quite distinct philosophies that are used for drawing conclusions with conventional statistical methods. One is design-based, drawing its justification from the randomization used in sampling, or in the random allocation of experimental units to different treatments. The other is model-based, drawing its justification from the random variation inherent in a model assumed to describe the nature of observations. In general it is recommended that where possible inferences should be design-based because this requires fewer assumptions and is always valid providing that randomizations are properly carried out.

- There are limitations with tests of significance which has led to their use being criticised, at least with some applications. Two particular problems are: (1) often tests are carried out when the null hypothesis is known to probably be untrue so that a significant result is very likely if enough data are collected, and (2) a non-significant result does not mean that the null hypothesis is false because the sample size may just not be large enough to detect the existing effect.

- It is argued that null hypotheses are relevant in situations where there really is doubt about whether a null hypothesis is true or not. If this is not the case, then it is more reasonable to estimate the magnitude of effects with some measure of how accurate the estimation is, using a confidence interval, for example.

- Randomization tests have been used quite often with environmental data. The idea is to compare the value of an observed test statistic with the distribution obtained by randomly reallocating the data to different samples in some sense. These tests have the advantage of requiring fewer assumptions than more conventional tests. They are, however, computer-intensive and may need special computer software.

- Bootstrapping is another computer-intensive method. It is based on the idea that in the absence of any knowledge about a distribution other than the values in a random sample from the distribution, the best guide to what would happen by resampling the population is to resample the sample. In principle, bootstrapping can be used to conduct tests of significance and to construct confidence intervals for population parameters.

- Pseudoreplication occurs when standard statistical methods are used to test treatment effects where either treatments are not replicated, or replicates are not statistically independent. Two errors can be made in this respect: (1) inferences can be extended outside the population actually sampled, and (2) observations that are correlated because they are close in space or time are not analysed taking this into account.

- When several related hypothesis tests are carried out at the same time and all the null hypotheses are true, the probability of at least one significant result increases with the number of tests. This is a well-known problem which has led to the development of a range of procedures to take into account the multiple testing. Multiple comparison methods to compare means at different factor levels after analysis of variance are procedures of this type that are widely available in statistical software, but not favoured by all statisticians. More general approaches are also used, including using a Bonferroni adjustment for the significance level used with individual tests, and Holm's stepwise method for adjusting these levels.

- Meta-analysis is concerned with the problem of combining the results from a number of different studies on the same subject. This can be done by combining the p-values obtained from different studies using Fisher's method, or the Stouffer method. The Stouffer method also has a variation called the Liptak-Stouffer method which allows the results of different studies to receive different weights. Alternatively, rather than using p-values, an effect size is estimated for each study and these effect sizes are examined in terms of the overall effect, the extent to which the effect varies from study to study, and whether the variation between studies can be explained by differences between the types of studies used.

- Bayesian inference is different from conventional statistical inference, and is becoming more widely used. With this approach a prior distribution assumed for a parameter of interest is changed using Bayes' theorem into a posterior distribution, given the information from some new data. Modern computing methods, particularly Markov chain Monte Carlo, make the Bayesian approach much easier to use than was the case in the past. However, it is cautioned that Bayesian inference is very much model-based, with all the potential problems that this implies.

CHAPTER 5

Environmental Monitoring

5.1 Introduction

The increasing worldwide concern about threats to the natural environment both on a local and a global scale has led to the introduction of many monitoring schemes that are intended to provide an early warning of violations of quality control systems, to detect the effects of major events such as accidental oil spills or the illegal disposal of wastes, and to study long-term trends or cycles in key environmental variables.

Examples of some of the national monitoring schemes that are now operating are the United States Environmental Protection Agency's Environmental Monitoring and Assessment Program (EMAP) based on 12,600 hexagons each with an area of 40 square kilometres, the United Kingdom Environmental Change Network (ECN) based on nine sites, and the Swedish Environmental Monitoring Program based on 20 sites. In all three of these schemes a large number of variables are recorded on a regular basis to describe physical aspects of the land, water and atmosphere, and the abundance of many species of animals and plants. Around the world numerous smaller scale monitoring schemes are also operated for particular purposes, such as to ensure that the quality of drinking water is adequate.

Monitoring schemes to detect unexpected changes and trends are essentially repeated surveys. The sampling methods described in Chapter 2 are therefore immediately relevant. In particular, if the mean value of a variable for the sample units in a geographical area is of interest, then the population of units should be randomly sampled so that the accuracy of estimates can be assessed in the usual way. Modifications of simple random sampling such as stratified sampling may well be useful to improve efficiency.

The requirements of environmental monitoring schemes have led to an interest in special types of sampling designs that include aspects of random sampling, good spatial cover, and the gradual replacement of sampling sites over time (Skalski, 1990; Stevens and Olsen, 1991; Overton *et al.*, 1991; Urquhart *et al.*, 1993; Conquest and Ralph, 1998). Designs that are optimum in some sense have also been developed (Fedorov and Mueller, 1989; Caselton *et al.*, 1992).

Although monitoring schemes sometimes require fairly complicated designs, as a general rule it is a good idea to keep designs as simple as possible so that they are easily understood by administrators and the public. Simple designs also make it easier to use the data for purposes that were not foreseen in the first place, which is something that will often occur. As noted by Overton and Stehman (1995, 1996), complex sample structures create potential serious difficulties that do not exist with simple random sampling.

5.2 Purposely Chosen Monitoring Sites

For practical reasons the sites for long-term monitoring programs are often not randomly chosen. For example, Cormack (1994) notes that the nine sites for the United Kingdom ECN were chosen on the basis of having:

(a) a good geographical distribution covering a wide range of environmental conditions and the principal natural and managed ecosystems;

(b) some guarantee of long-term physical and financial security;

(c) a known history of consistent management;

(d) reliable and accessible records of past data, preferably for ten or more years; and

(e) sufficient size to allow the opportunity for further experiments and observations.

In this scheme it is assumed that the initial status of sites can be allowed for by only considering time changes. These changes can then be related to differences between the sites in terms of measured meteorological variables and known geographical differences.

5.3 Two Special Monitoring Designs

Skalski (1990) suggested a rotating panel design with augmentation for long-term monitoring. This takes the form shown in Table 5.1 if there are eight sites that are visited every year and four sets of ten sites that are rotated. Site set 7, for example, consists of ten sites that are visited in years 4 to 7 of the study. The number of sites in different sets is arbitrary. Preferably, the sites will be randomly chosen from an appropriate population of sites. This design has some appealing properties: the sites that are always measured can be used to detect long-term trends but the rotation of blocks of ten sites ensures that the study is not too dependent on an initial choice of sites that may be unusual in some respects.

Table 5.1 Skalski's (1990) rotating panel design with augmentation. Every year 48 sites are visited. Of these, 8 are always the same and the other 40 sites are in four blocks of size ten, such that each block of ten remains in the sample for four years after the initial start up period

Site set	Number of sites	1	2	3	4	5	6	7	8	9	10	11	12
0	8	x	x	x	x	x	x	x	x	x	x	x	x
1	10	x											
2	10	x	x										
3	10	x	x	x									
4	10	x	x	x	x								
5	10		x	x	x	x							
6	10			x	x	x	x						
7	10				x	x	x	x					
.	.												
.	.												
.	.												
14	10											x	x
15	10												x

The serially alternating design with augmentation that is used for EMAP is of the form shown in Table 5.2. It differs because sites are not rotated out of the study. Rather, there are eight sites that are measured every year and another 160 sites in blocks of 40, where each block of 40 is measured every four years. The number of sites in different sets is a choice in a design of this form. Sites should be randomly selected from an appropriate population.

Urquhart *et al.* (1993) compared the efficiency of the designs in Tables 5.1 and 5.2 when there are a total of 48 sites, of which the number visited every year (i.e., in set 0) ranged from 0 to 48. To do this, they assumed the model

$$Y_{ijk} = S_{i(j)k} + T_j + e_{ijk},$$

where Y_{ijk} is a measure of the condition at site i, in year j, within site set k; $S_{i(j)k}$ is an effect specific to site i, in site set k, in year j, T_j is a year effect common to all sites, and e_{ijk} is a random disturbance. They also allowed for autocorrelation between the overall year effects, and between the repeated measurements at one site. They found the design of Table 5.2 to always be better for estimating the current mean and the slope in a trend because more sites are measured in the first few years of the study. However, in a more recent study which compared the two designs in terms of variance and cost, Lesser and Kalsbeek (1997) concluded that the first design tends to be better for detecting short-term change while the second design tends to be better for detecting long-term change.

Table 5.2 A serially alternating design with augmentation. Every year 48 sites are measured. Of these, eight sites are always the same and the other 40 sites are measured every four years

Site set	Number of sites	1	2	3	4	5	6	7	8	9	10	11	12
0	8	x	x	x	x	x	x	x	x	x	x	x	x
1	40	x				x				x			
2	40		x				x				x		
3	40			x				x				x	
4	40				x				x				x

The EMAP sample design is based on approximately 12,600 points on a grid, each of which is the centre of a hexagon with area 40 km². The grid is itself within a large hexagonal region covering much of North America, as shown in Figure 5.1. The area covered by the 40 km² hexagons entered on the grid points is one sixteenth of the total area of the conterminous United States, with the area used being chosen after a random shift in the grid. Another aspect of the design is that the four sets of sites that are measured on different years are spatially interpenetrating, as indicated in Figure 5.2. This allows the estimation of parameters for the whole area every year.

Figure 5.1 The EMAP baseline grid for North America. The shaded area shown is covered by about 12,600 small hexagons, with a spacing between their centres being of 27 km.

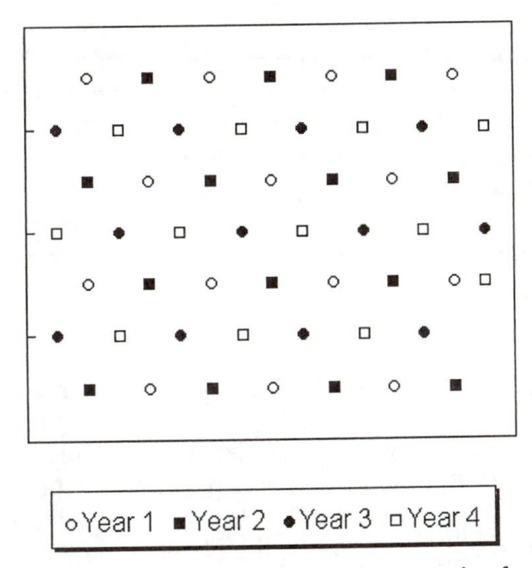

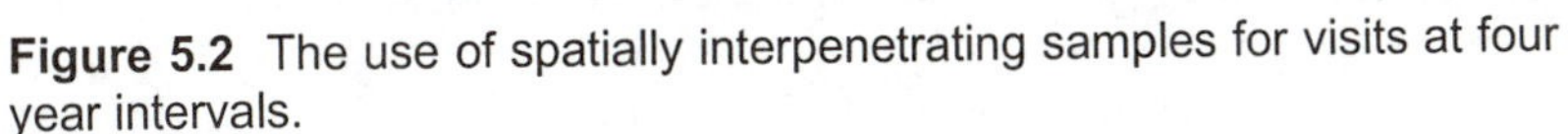

Figure 5.2 The use of spatially interpenetrating samples for visits at four year intervals.

5.4 Designs Based on Optimization

One approach to the design of monitoring schemes is by choosing the sites so that the amount of information is in some sense maximized. The main question then is how to measure the information that is to be maximized, particularly if the monitoring scheme has a number of different objectives, some of which will only become known in the future.

One possibility involves choosing a network design, or adding or subtracting stations to minimize entropy, where low entropy corresponds to high 'information' (Caselton *et al.*, 1992). The theory is complex, and needs more prior information than will usually be available, particularly if there is no existing network to provide this.

Another possibility considers the choice of a network design to be a problem of the estimation of a regression function for which a classical theory of optimal design exists (Fedorov and Mueller, 1989).

5.5 Monitoring Designs Typically Used

In practice, sample designs for monitoring often consist of selecting a certain number of sites preferably (but not necessarily) at random from the potential sites in a region, and then measuring the variable of interest at those sites at a number of points in time. A complication is that for one reason or another some of the sites may not be measured at some of the times. A typical set of data will then look like the data in Table 5.3 for pH values measured on lakes in Norway. With this set of data, which is part of the more extensive data that are shown in Table 1.1 and discussed in Example 1.2, the main question of interest is whether there is any evidence for changes from year to year in the general level of pH and, in particular, whether the pH level was tending to increase or decrease.

5.6 Detection of Changes by Analysis of Variance

A relatively simple analysis for data like the Norwegian lake pH values shown in Table 5.3 involves carrying out a two factor analysis of variance, as discussed in Section 3.5. The two factors are then the site and the time. The model for the observation at site i at time j is

$$y_{ij} = \mu + S_i + T_j + e_{ij}, \qquad (5.1)$$

where μ represents an overall general level for the variable being measured, S_i represents the deviation of site i from the general level, T_j represents a time effect, and e_{ij} represents measurement errors and other random variation that is associated with the observation at the site at the particular time.

The model (5.1) does not include a term for the interaction between sites and times as is included in the general two-factor analysis of variance model as defined in equation (4.31). This is because there is only at most one observation for a site in a particular year, which means that it is not possible to separate interactions from measurement errors. Consequently, it must be assumed that any interactions are negligible.

Example 5.1 Analysis of Variance on the pH Values

The results of an analysis of variance on the pH values for Norwegian lakes are summarised in Table 5.4. The results in this table were obtained using the MINITAB package (Minitab Inc., 1994) using an option that takes into account the missing values, although many other standard statistical packages could have been used just as well. The effects in the model were assumed to be fixed rather than random (as discussed in Section 4.5), although since interactions are assumed to be negligible the same results would be obtained using random effects. It is found that there is a significant difference between the lakes ($p = 0.000$) and a nearly significant difference between the years ($p = 0.061$). Therefore there is no very strong evidence from this analysis of differences between years.

To check the assumptions of the analysis, standardized residuals (the differences between the actual observations and those predicted by the model, divided by their standard deviations) can be plotted against the lake, the year, and against their position in space for each of the four years. These plots are shown in Figures 5.3 and 5.4. These residuals show no obvious patterns so that the model seems satisfactory, except that there are one or two residuals that are rather large.

Table 5.3 Values for pH for lakes in southern Norway with the latitudes (Lat) and longitudes (Long) for the lakes

			pH			
Lake	Lat	Long	1976	1977	1978	1981
1	58.0	7.2	4.59		4.48	4.63
2	58.1	6.3	4.97		4.60	4.96
4	58.5	7.9	4.32	4.23	4.40	4.49
5	58.6	8.9	4.97	4.74	4.98	5.21
6	58.7	7.6	4.58	4.55	4.57	4.69
7	59.1	6.5	4.80		4.74	4.94
8	58.9	7.3	4.72	4.81	4.83	4.90
9	59.1	8.5	4.53	4.70	4.64	4.54
10	58.9	9.3	4.96	5.35	5.54	5.75
11	59.4	6.4	5.31	5.14	4.91	5.43
12	58.8	7.5	5.42	5.15	5.23	5.19
13	59.3	7.6	5.72		5.73	5.70
15	59.3	9.8	5.47		5.38	5.38
17	59.1	11.8	4.87	4.76	4.87	4.90
18	59.7	6.2	5.87	5.95	5.59	6.02
19	59.7	7.3	6.27	6.28	6.17	6.25
20	59.9	8.3	6.67	6.44	6.28	6.67
21	59.8	8.9	6.06		5.80	6.09
24	60.1	12.0	5.38	5.32	5.33	5.21
26	59.6	5.9	5.41	5.94		
30	60.4	10.2	5.60	6.10	5.57	5.98
32	60.4	12.2	4.93	4.94	4.91	4.93
34-1	60.5	5.5			4.90	4.87
36	60.9	7.3	5.60	5.69	5.41	5.66
38	60.9	10.0	6.72	6.59	6.39	
40	60.7	12.2	5.97	6.02	5.71	5.67
41	61.0	5.0	4.68	4.72	5.02	
42	61.3	5.6	5.07			5.18
43	61.0	6.9	6.23	6.34	6.20	6.29
46	61.0	9.7	6.64		6.24	6.37
47	61.3	10.8	6.15	6.23	6.07	5.68
49	61.5	4.9	4.82	4.77	5.09	5.45
50	61.5	5.5	5.42	4.82	5.34	5.54
57	61.7	4.9	4.99		5.16	5.25
58	61.7	5.8	5.31	5.77	5.60	5.55
59	61.9	7.1	6.26	5.03	5.85	
65	62.2	6.4	5.99	6.10	5.99	6.13
80	58.1	6.7	4.63		4.59	4.92
81	58.3	8.0	4.47		4.36	4.50
82	58.7	7.1	4.60		4.54	4.66
83	58.9	6.1	4.88	4.99	4.86	4.92
85	59.4	11.3	4.60	4.88	4.91	4.84
86	59.3	9.4	4.85	4.65	4.77	4.84
87	59.2	7.6	5.06		5.15	5.11
88	59.4	7.3	5.97	5.82	5.90	6.17
89	59.3	6.3	5.47		6.05	5.82
94	61.0	11.5	6.05	5.97	5.78	5.75
95-1	61.2	4.6			5.70	5.50
Mean			5.34	5.40	5.31	5.38
SD			0.65	0.66	0.57	0.56

Table 5.4 Analysis of variance table for the data on pH levels in Norwegian lakes

Source of Variation	Sum of Squares[1]	Degrees of Freedom	Mean Square	F	Significance level (p)
Lake	58.70	47	1.249	37.95	0.000
Year	0.25	3	0.083	2.53	0.061
Error	3.85	117	0.033		
Total	62.80	167			

[1]The sums of squares shown here depend on the order in which effects are added into the model, which is species, then the treatment, and finally the interaction between these two factors.

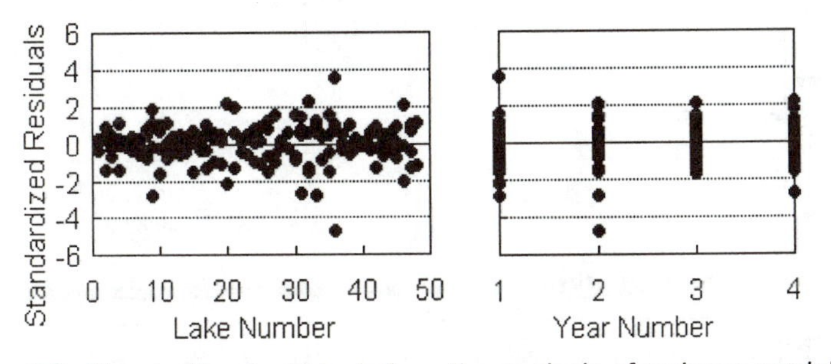

Figure 5.3 Standardized residuals from the analysis of variance model for pH in Norwegian lakes plotted against the lake number and the year number.

5.7 Detection of Changes Using Control Charts

Control charts are used to monitor industrial processes (Montgomery, 1991) and they can be used equally well with environmental data. The simplest approach involves using an $\bar{x}$ chart to detect changes in a process mean, together with a range chart to detect changes in the amount of variation. These types of charts are often called Shewhart control charts after their originator (Shewhart, 1931).

Typically, the starting point is a moderately large set of data consisting of M random samples of size n, where these are taken at equally spaced intervals of time from the output of the process. This set of data is then used to estimate the process mean and standard deviation, and hence to construct the two charts. The data are then plotted on the charts. It is usually assumed that the observations are normally distributed.

If the process seems to have a constant mean and standard deviation, then the sampling of the process is continued with new points being plotted to monitor whatever is being measured. If the mean or standard deviation does not seem to have been constant for the time when the initial samples were taken, then in the industrial process situation, action is taken to bring the process under control. With environmental monitoring this may not be possible. However, the knowledge that the process being measured is not stable will be of interest anyway.

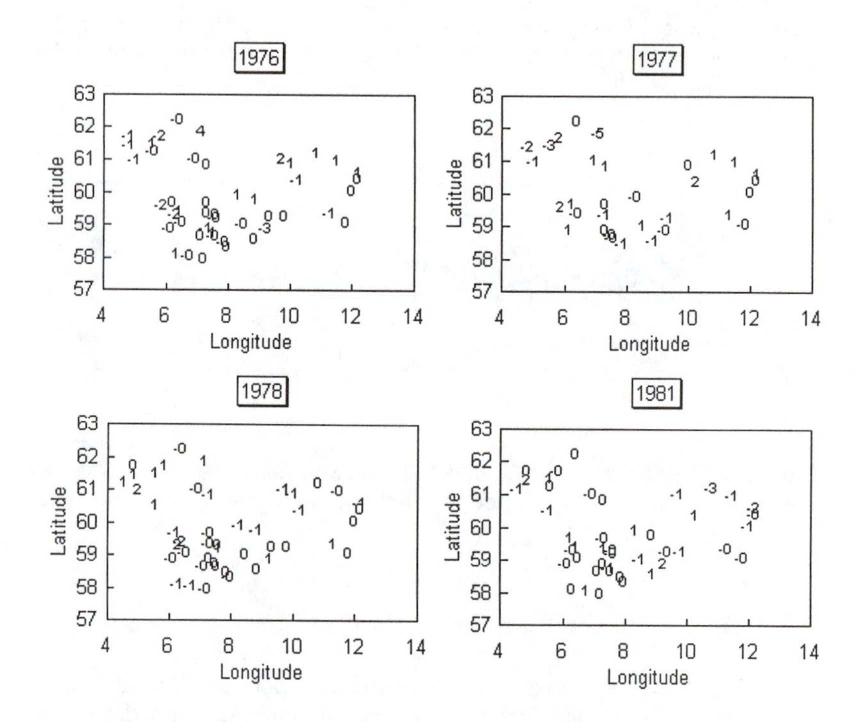

Figure 5.4 Standardized residuals from the analysis of variance model for pH in Norwegian lakes plotted against the locations of the lakes. The standardized residuals are rounded to the nearest integer for clarity.

The method for constructing the $\bar{x}$-chart involves the following stages:

1. The sample mean and the sample range (the maximum value in a sample minus the minimum value in a sample) are calculated, for each of the M samples. For the ith sample let these values be denoted by $\bar{x}_i$ and R_i.

2. The mean of the variable being measured is assumed to be constant, and is estimated by the overall mean of all the available observations, which is also just the mean of the sample means $\bar{x}_1$ to $\bar{x}_M$. Let the estimated mean be denoted by $\hat{\mu}$.

3. Similarly, the standard deviation of ring widths is assumed to have remained constant and this is estimated on the basis of a known relationship between the mean range for samples of size n and the standard deviation for samples from a normal distribution. This relationship is of the form $\sigma = k(n)\mu_R$, where μ_R is the mean range for samples of size n, and the constant k(n) is given in Table 5.5. Thus the estimated standard deviation is

$$\hat{\sigma} = k(n)\bar{R}, \tag{5.2}$$

   where $\bar{R}$ is the mean of the sample ranges.

4. The standard error of the mean for samples of size n is estimated to be $S\hat{E}(\bar{x}) = \hat{\sigma}/\sqrt{n}$.

5. Warning limits are set at the mean plus and minus 1.96 standard errors, i.e., at $\hat{\mu} \pm 1.96S\hat{E}(\bar{x})$. If the mean and standard deviation are constant then only about 1 in 20 (5%) sample means should be outside one of these limits. Action limits are set at the mean plus and minus 3.09 standard errors, i.e., at $\hat{\mu} \pm 3.09S\hat{E}(\bar{x})$. Only about 1 in 500 (0.2%) sample means should plot outside these limits.

The rationale behind constructing the $\bar{x}$ chart in this way is that it shows the changes in the sample means with time, and the warning and action limits indicate whether these changes are too large to be due to normal random variation if the mean is in fact constant.

Table 5.5 Control chart limits for sample ranges, assuming samples from normal distributions. To find the limits on the range chart, multiply the mean range by the tabulated value. For example, for samples of size n = 5 the lower action limit is $0.16\mu_R$, where μ_R is the mean range. With a stable distribution a warning limit is crossed with probability 0.05 (5%) and an action limit with probability 0.002 (0.2%). The last column is the factor that the mean range must be multiplied by to obtain the standard deviation. For example, for samples of size 3 the standard deviation is $0.591\mu_R$. Source: Tables G1 and G2 of Davies and Goldsmith (1972)

Sample	Lower limits		Upper limits		SD
size	Action	Warning	Warning	Action	Factor (k)
2	0.00	0.04	2.81	4.12	0.887
3	0.04	0.18	2.17	2.99	0.591
4	0.10	0.29	1.93	2.58	0.486
5	0.16	0.37	1.81	2.36	0.430
6	0.21	0.42	1.72	2.22	0.395
7	0.26	0.46	1.66	2.12	0.370
8	0.29	0.50	1.62	2.04	0.351
9	0.32	0.52	1.58	1.99	0.337
10	0.35	0.54	1.56	1.94	0.325

With control charts it is conventional to measure process variability using sample ranges on the grounds of simplicity, although standard deviations or variances could be used instead. Like $\bar{x}$ charts, range charts can have warning limits placed so that the probability of crossing one of these is 0.05 (5%), assuming that the level of variation is stable. Similarly, action limits can be placed so that the probability of crossing one of them is 0.002 (0.2%) when the level of variation is stable. The setting of these limits requires the use of tabulated values that are provided and explained in Table 5.5.

Control charts can be produced quite easily in a spreadsheet program. Alternatively, statistical package such as MINITAB (Mintab Inc., 1994) have options to produce the charts, and often allow a number of other types of control charts to be produced as well.

Example 5.2 Monitoring pH in New Zealand

Table 5.6 shows some data that were obtained from regular monitoring of rivers in the South Island of New Zealand. Values are provided for pH, for five randomly chosen rivers, with a different selection for each of the monthly sample times from January 1989 to December 1997. The data are used to construct control charts for monitoring pH over the sampled time. As shown in Figure 5.5, the distribution is reasonably close to normal.

The overall mean of the pH values for all the samples is $\hat{\mu} = 7.640$. This is used as the best estimate of the process mean. The mean of the sample ranges is $\bar{R} = 0.694$. From Table 5.5, the factor to convert this to an estimate of the process standard deviation is k(5) = 0.43. The estimated standard deviation is therefore

$$\hat{\sigma} = 0.43 \times 0.694 = 0.298.$$

Hence, the estimated standard error for the sample means is

$$\hat{SE}(\bar{x}) = 0.298/\sqrt{5} = 0.133.$$

The mean control chart is shown in Figure 5.6 (a), with the action limits set at 7.640 ± 3.09 x 0.133 (i.e., 7.23 and 8.05), and the warning limits at 7.640 ± 1.96 x 0.133 (i.e., 7.38 and 7.90).

The range chart is shown in Figure 5.6 (b). Using the factors for setting the action and warning limits from Table 5.5, these limits are at are 0.16 x 0.694 = 0.11 (lower action limit), 0.37 x 0.694 = 0.26 (lower warning limit), 1.81 x 0.694 = 1.26 (upper warning limit), and 2.36 x 0.694 = 1.64 (upper action limit). Due to the nature of the distribution of sample ranges, these limits are not symmetrically placed about the mean.

With 108 observations altogether, it is expected that about five points will plot outside the warning limits. In fact, there are ten points outside these limits, and one point outside the lower action limit. It appears, therefore, that the mean pH level in the South Island of New Zealand was changing to some extent during the monitored period, with the overall plot suggesting that the pH level was high for about the first two years, and was lower from then on.

The range chart has seven points outside the warning limits, and one point above the upper action limit. Here again there is evidence that the process variation was not constant, with occasional 'spikes' of high variability, particularly in the early part of the monitoring period.

Table 5.6 Values for pH for five randomly selected rivers in the South Island of New Zealand, for each month from January 1989 to December 1997. This is part of a larger data set provided by Graham McBride, National Institute of Water and Atmospheric Research, Hamilton, New Zealand

Year	Month	pH Values					Mean	Range
1989	Jan	7.27	7.10	7.02	7.23	8.08	7.34	1.06
	Feb	8.04	7.74	7.48	8.10	7.21	7.71	0.89
	Mar	7.50	7.40	8.33	7.17	7.95	7.67	1.16
	Apr	7.87	8.10	8.13	7.72	7.61	7.89	0.52
	May	7.60	8.46	7.80	7.71	7.48	7.81	0.98
	Jun	7.41	7.32	7.42	7.82	7.80	7.55	0.50
	Jul	7.88	7.50	7.45	8.29	7.45	7.71	0.84
	Aug	7.88	7.79	7.40	7.62	7.47	7.63	0.48
	Sep	7.78	7.73	7.53	7.88	8.03	7.79	0.50
	Oct	7.14	7.96	7.51	8.19	7.70	7.70	1.05
	Nov	8.07	7.99	7.32	7.32	7.63	7.67	0.75
	Dec	7.21	7.72	7.73	7.91	7.79	7.67	0.70
1990	Jan	7.66	8.08	7.94	7.51	7.71	7.78	0.57
	Feb	7.71	8.73	8.18	7.04	7.28	7.79	1.69
	Mar	7.72	7.49	7.62	8.13	7.78	7.75	0.64
	Apr	7.84	7.67	7.81	7.81	7.80	7.79	0.17
	May	8.17	7.23	7.09	7.75	7.40	7.53	1.08
	Jun	7.79	7.46	7.13	7.83	7.77	7.60	0.70
	Jul	7.16	8.44	7.94	8.05	7.70	7.86	1.28
	Aug	7.74	8.13	7.82	7.75	7.80	7.85	0.39
	Sep	8.09	8.09	7.51	7.97	7.94	7.92	0.58
	Oct	7.20	7.65	7.13	7.60	7.68	7.45	0.55
	Nov	7.81	7.25	7.80	7.62	7.75	7.65	0.56
	Dec	7.73	7.58	7.30	7.78	7.11	7.50	0.67
1991	Jan	8.52	7.22	7.91	7.16	7.87	7.74	1.36
	Feb	7.13	7.97	7.63	7.68	7.90	7.66	0.84
	Mar	7.22	7.80	7.69	7.26	7.94	7.58	0.72
	Apr	7.62	7.80	7.59	7.37	7.97	7.67	0.60
	May	7.70	7.07	7.26	7.82	7.51	7.47	0.75
	Jun	7.66	7.83	7.74	7.29	7.30	7.56	0.54
	Jul	7.97	7.55	7.68	8.11	8.01	7.86	0.56
	Aug	7.86	7.13	7.32	7.75	7.08	7.43	0.78
	Sep	7.43	7.61	7.85	7.77	7.14	7.56	0.71
	Oct	7.77	7.83	7.77	7.54	7.74	7.73	0.29
	Nov	7.84	7.23	7.64	7.42	7.73	7.57	0.61
	Dec	8.23	8.08	7.89	7.71	7.95	7.97	0.52

Table 5.6 (Continued)

Year	Month	pH Values					Mean	Range
1992	Jan	8.28	7.96	7.86	7.65	7.49	7.85	0.79
	Feb	7.23	7.11	8.53	7.53	7.78	7.64	1.42
	Mar	7.68	7.68	7.15	7.68	7.85	7.61	0.70
	Apr	7.87	7.20	7.42	7.45	7.96	7.58	0.76
	May	7.94	7.35	7.68	7.50	7.12	7.52	0.82
	Jun	7.80	6.96	7.56	7.22	7.76	7.46	0.84
	Jul	7.39	7.12	7.70	7.47	7.74	7.48	0.62
	Aug	7.42	7.41	7.47	7.80	7.12	7.44	0.68
	Sep	7.91	7.77	6.96	8.03	7.24	7.58	1.07
	Oct	7.59	7.41	7.41	7.02	7.60	7.41	0.58
	Nov	7.94	7.32	7.65	7.84	7.86	7.72	0.62
	Dec	7.64	7.74	7.95	7.83	7.96	7.82	0.32
1993	Jan	7.55	8.01	7.37	7.83	7.51	7.65	0.64
	Feb	7.30	7.39	7.03	8.05	7.59	7.47	1.02
	Mar	7.80	7.17	7.97	7.58	7.13	7.53	0.84
	Apr	7.92	8.22	7.64	7.97	7.18	7.79	1.04
	May	7.70	7.80	7.28	7.61	8.12	7.70	0.84
	Jun	7.76	7.41	7.79	7.89	7.36	7.64	0.53
	Jul	8.28	7.75	7.76	7.89	7.82	7.90	0.53
	Aug	7.58	7.84	7.71	7.27	7.95	7.67	0.68
	Sep	7.56	7.92	7.43	7.72	7.21	7.57	0.71
	Oct	7.19	7.73	7.21	7.49	7.33	7.39	0.54
	Nov	7.60	7.49	7.86	7.86	7.80	7.72	0.37
	Dec	7.50	7.86	7.83	7.58	7.45	7.64	0.41
1994	Jan	8.13	8.09	8.01	7.76	7.24	7.85	0.89
	Feb	7.23	7.89	7.81	8.12	7.83	7.78	0.89
	Mar	7.08	7.92	7.68	7.70	7.40	7.56	0.84
	Apr	7.55	7.50	7.52	7.64	7.14	7.47	0.50
	May	7.75	7.57	7.44	7.61	8.01	7.68	0.57
	Jun	6.94	7.37	6.93	7.03	6.96	7.05	0.44
	Jul	7.46	7.14	7.26	6.99	7.47	7.26	0.48
	Aug	7.62	7.58	7.09	6.99	7.06	7.27	0.63
	Sep	7.45	7.65	7.78	7.73	7.31	7.58	0.47
	Oct	7.65	7.63	7.98	8.06	7.51	7.77	0.55
	Nov	7.85	7.70	7.62	7.96	7.13	7.65	0.83
	Dec	7.56	7.74	7.80	7.41	7.59	7.62	0.39
1995	Jan	8.18	7.80	7.22	7.95	7.79	7.79	0.96
	Feb	7.63	7.88	7.90	7.45	7.97	7.77	0.52
	Mar	7.59	8.06	8.22	7.57	7.73	7.83	0.65
	Apr	7.47	7.82	7.58	8.03	8.19	7.82	0.72
	May	7.52	7.42	7.76	7.66	7.76	7.62	0.34
	Jun	7.61	7.72	7.56	7.49	6.87	7.45	0.85
	Jul	7.30	7.90	7.57	7.76	7.72	7.65	0.60
	Aug	7.75	7.75	7.52	8.12	7.75	7.78	0.60
	Sep	7.77	7.78	7.75	7.49	7.14	7.59	0.64
	Oct	7.79	7.30	7.83	7.09	7.09	7.42	0.74
	Nov	7.87	7.89	7.35	7.56	7.99	7.73	0.64
	Dec	8.01	7.56	7.67	7.82	7.44	7.70	0.57

Table 5.6 (Continued)

Year	Month	pH Values					Mean	Range
1996	Jan	7.29	7.62	7.95	7.72	7.98	7.71	0.69
	Feb	7.50	7.50	7.90	7.12	7.69	7.54	0.78
	Mar	8.12	7.71	7.20	7.43	7.56	7.60	0.92
	Apr	7.64	7.75	7.80	7.72	7.73	7.73	0.16
	May	7.59	7.57	7.86	7.92	7.22	7.63	0.70
	Jun	7.60	7.97	7.14	7.72	7.72	7.63	0.83
	Jul	7.07	7.70	7.33	7.41	7.26	7.35	0.63
	Aug	7.65	7.68	7.99	7.17	7.72	7.64	0.82
	Sep	7.51	7.64	7.25	7.82	7.91	7.63	0.66
	Oct	7.81	7.53	7.88	7.11	7.50	7.57	0.77
	Nov	7.16	7.85	7.63	7.88	7.66	7.64	0.72
	Dec	7.67	8.05	8.12	7.38	7.77	7.80	0.74
1997	Jan	7.97	7.04	7.48	7.88	8.24	7.72	1.20
	Feb	7.17	7.69	8.15	6.96	7.47	7.49	1.19
	Mar	7.52	7.84	8.12	7.85	8.07	7.88	0.60
	Apr	7.65	7.14	7.38	7.23	7.66	7.41	0.52
	May	7.62	7.64	8.17	7.56	7.53	7.70	0.64
	Jun	7.10	7.16	7.71	7.57	7.15	7.34	0.61
	Jul	7.85	7.62	7.68	7.71	7.72	7.72	0.23
	Aug	7.39	7.53	7.11	7.39	7.03	7.29	0.50
	Sep	8.18	7.75	7.86	7.77	7.77	7.87	0.43
	Oct	7.67	7.66	7.87	7.82	7.51	7.71	0.36
	Nov	7.57	7.92	7.72	7.73	7.47	7.68	0.45
	Dec	7.97	8.16	7.70	8.21	7.74	7.96	0.51
						Mean	7.640	0.694

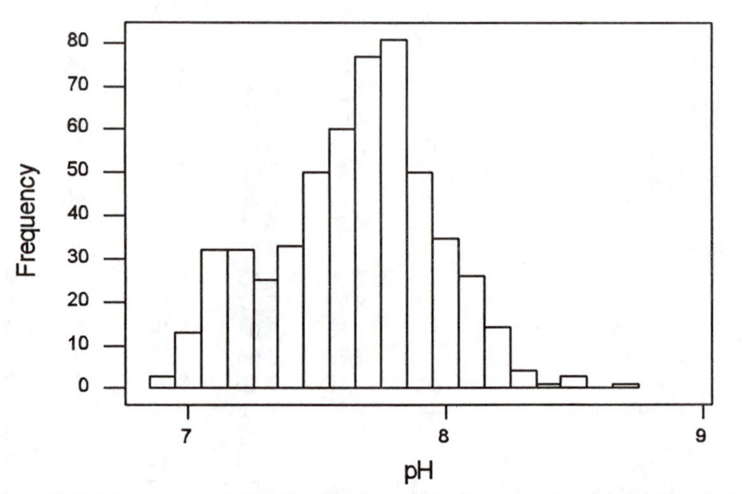

Figure 5.5 Histogram of the distribution of pH for samples from lakes in the South Island of New Zealand, 1989 to 1997.

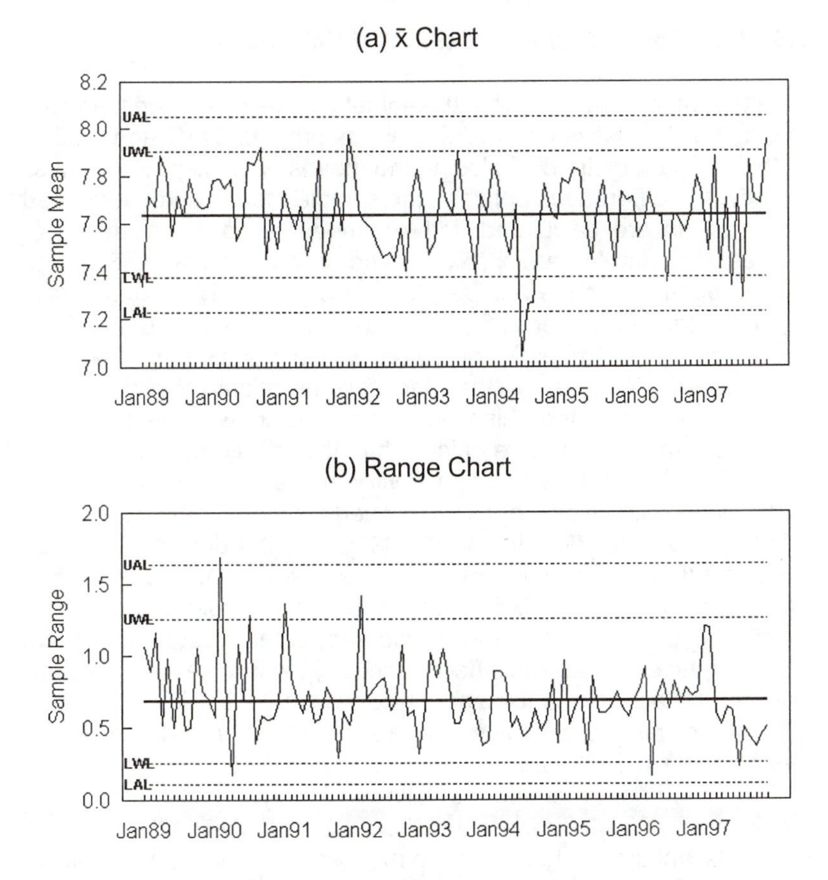

Figure 5.6 Control charts for pH levels in rivers in the South Island of New Zealand. For a process that is stable, about 1 in 40 points should plot outside one of the warning limits (LWL and UWL) and only about 1 in 1000 points should plot outside one of the action limits (LAL and UAL).

As this example demonstrates, control charts are a relatively simple tool for getting some insight into the nature of the process being monitored, through the use of the action and control limits, which indicate the level of variation expected if the process is stable. See a book on industrial process control such as the one by Montgomery (1991) for more details about the use of $\bar{x}$ and range charts, as well as a number of other types of charts that are intended to detect changes in the average level and the variation in processes.

5.8 Detection of Changes Using CUSUM Charts

An alternative to the $\bar{x}$ chart that is sometimes used for monitoring the mean of a process is a CUSUM chart, as proposed by Page (1961). With this chart, instead of plotting the means of successive samples directly, the differences between the sample means and the desired target mean are calculated and summed. Thus if the means of successive samples are $\bar{x}_1$, $\bar{x}_2$, $\bar{x}_3$, and so on, then $S_1 = (\bar{x}_1 - \mu)$ is plotted against sample number 1, $S_2 = (\bar{x}_1 - \mu) + (\bar{x}_2 - \mu)$ is plotted against sample number 2, $S_3 = (\bar{x}_1 - \mu) + (\bar{x}_2 - \mu) + (\bar{x}_3 - \mu)$ is plotted against sample number 3, and so on, where μ is the target mean of the process. The idea is that because deviations from the target mean are accumulated, this type of chart will show small deviations from the target mean more quickly than the traditional $\bar{x}$ chart. See MacNally and Hart (1997) for an environmental application.

Setting control limits for CUSUM charts is more complicated than it is for an $\bar{x}$ chart (Montgomery, 1991), and the details will not be considered here. There is, however, a variation on the usual CUSUM approach that will be discussed because of its potential value for monitoring under circumstances where repeated measurements are made on the same sites at different times. That is to say, the data are like the pH values for a fixed set of Norwegian lakes shown in Table 5.3, rather than the pH values for random samples of New Zealand rivers shown in Table 5.6.

This method proceeds as follows (Manly, 1994; Manly and MacKenzie, 2000). Suppose that there are n sample units measured at m different times. Let x_{ij} be the measurement on sample unit i at time t_j, and let $\bar{x}_i$ be the mean of all the measurements on the unit. Assume that the units are numbered in order of their values for $\bar{x}_i$, so that $\bar{x}_1$ is the smallest mean and $\bar{x}_n$ is the largest mean. Then it is possible to construct m CUSUM charts by calculating

$$S_{ij} = (x_{1j} - \bar{x}_1) + (x_{2j} - \bar{x}_2) + \ldots + (x_{ij} - \bar{x}_i), \qquad (5.3)$$

for j from 1 to m and i from 1 to n, and plotting the S_{ij} values against i, for each of the m sample times.

The CUSUM chart for time t_j indicates the manner in which the observations made on sites at that time differ from the average values for all sample times. For example a positive slope for the CUSUM shows that the values for that time period are higher than the average values for all periods. Furthermore, a CUSUM slope of D for a series of sites (the rise divided by the number of observations) indicates that the observations on those sites are on average D higher than the corresponding means for the sites over all sample times. Thus a constant difference between the values on a sample unit at one time and the mean for all times is indicated by a constant slope of the CUSUM going either up or down from left to right. On the other hand, a positive slope on the left-hand side of the graph followed by a negative slope on the right-hand side indicates that the values at the time being considered were high for sites with a low mean but low for sites with a high mean. Figure 5.7 illustrates some possible patterns that might be obtained, and their interpretation.

Randomization methods can be used to decide whether the CUSUM plot for time t_j indicates systematic differences between the data for this time and the average for all times. In brief, four approaches based on the null hypothesis that the values for each sample unit are in a random order with respect to time are:

(a) A large number of randomized CUSUM plots can be constructed, where for each one of these the observations on each sample unit are randomly permuted. Then, for each value of i, the maximum and minimum values obtained for S_{ij} can be plotted on the CUSUM chart. This gives an envelope within which any CUSUM plot for real data can be expected to lie, as shown in Figure 5.8. If the real data plot goes outside the envelope, then there is clear evidence that the null hypothesis is not true.

(b) Using the randomizations it is possible to determine whether S_{ij} is significantly different from 0 for any particular values of i and j. Thus if there are R-1 randomizations, then the observed value of $|S_{ij}|$ is significantly different from 0 at the 100 α% level if it is among the largest 100α% of the R values of $|S_{ij}|$ consisting of the observed value and the R-1 randomizations.

(c) To obtain a statistic that measures the overall extent to which a CUSUM plot differs from what is expected on the basis of the null hypothesis, the standardized deviation of S_{ij} from zero,

$$Z_{ij} = S_{ij}/\sqrt{Var(S_{ij})},$$

is calculated for each value of i. A measure of the overall deviation of the jth CUSUM from 0 is then

$$Z_{max,j} = Max(|Z_{1j}|, |Z_{2j}|, ..., |Z_{nj}|).$$

The value of $Z_{max,j}$ for the observed data is significantly large at the 100α% level if it is among the largest 100α% of the R values consisting of itself plus R-1 other values obtained by randomization.

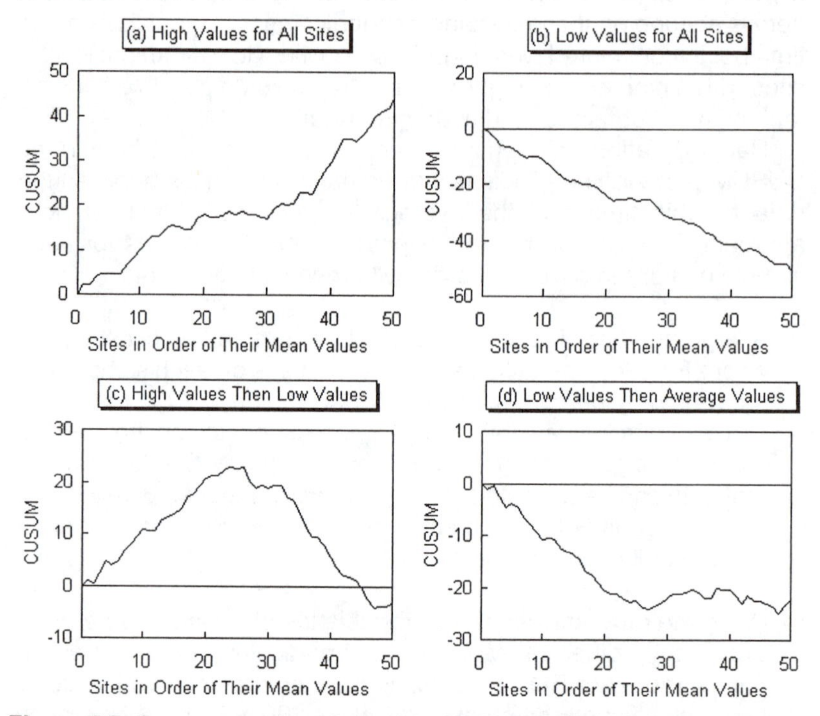

Figure 5.7 Some examples of possible CUSUM plots for a particular time period: (a) all the sites tend to have high values for this period; (b) all the sites tend to have low values for this period; (c) sites with a low mean for all periods tend to have relatively high values for this period, but sites with a high mean for all periods tend to have low values for this period; and (d) sites with a low mean for all periods tend to have even lower low values in this period, but sites with a high mean for all periods tend to have values close to this mean for this period.

(d) To measure the extent to which the CUSUM plots for all times differ from what is expected from the null hypothesis the statistic

$$Z_{max,T} = Z_{max,1} + Z_{max,2} + \dots + Z_{max,m}$$

can be used. Again, the $Z_{max,T}$ for a set of observed data is significantly large at the 100α% level if it is among the largest

$100\alpha\%$ of the R values consisting of itself plus R-1 comparable values obtained from randomized data.

See Manly (1994) and Manly and MacKenzie (2000) for more details about how these tests are made.

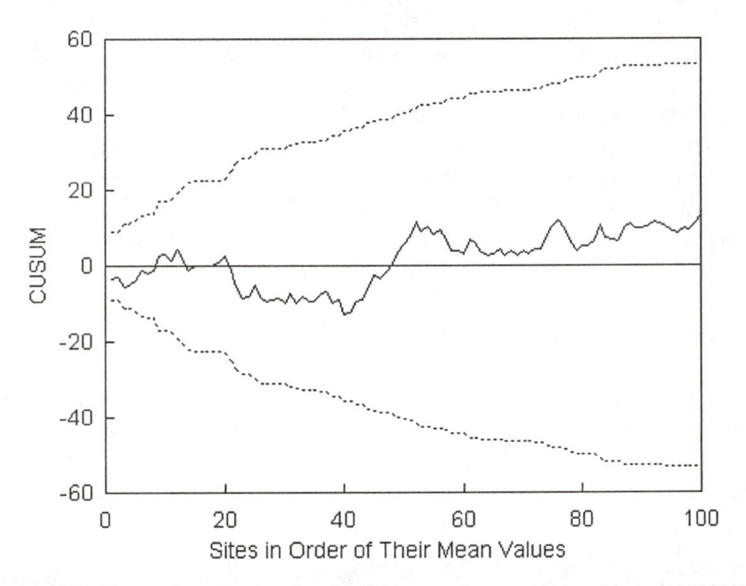

Figure 5.8 Example of a randomization envelope obtained for a CUSUM plot. For each site in order the minimum and maximum (as indicated by broken lines) are those obtained from all the randomized data sets for the order number being considered. For example, with 1000 randomizations, all of the CUSUM plots obtained from the randomized data are contained within the envelope. In this example the observed CUSUM plot is well within the randomization limits.

Missing values can be handled easily. If x_{ij} is not measured then it can be set equal to $\bar{x}_i$ for calculating the CUSUM using equation (5.3), where $\bar{x}_i$ is the mean of the observations that are present, while in equation (5.4) $Var(d_k)$ can be modified to equal the variance for the observations that are present. For the randomization tests only the observations that are present are randomly permuted on a sample unit.

The way that the CUSUM values are calculated using equation (5.3) means that any average differences between sites are allowed for, so that there is no requirement for the mean values at different sites to be similar. However, for the randomization tests there is an implicit assumption that there is negligible serial correlation between the successive observations at one sampling site, i.e., there is no tendency for these observations to be similar because they are relatively close in time. If this is not the case, as may well happen for many monitored variables, then it is possible to modify the randomization tests to allow for this. A method to do this is described by Manly and MacKenzie (2000), who also compare the power of the CUSUM procedure with the power of other tests for systematic changes in the distribution of the variable being studied. This modification will not be considered here.

The calculations for the CUSUM method are fairly complicated, particularly when serial correlation is involved. A Windows computer program called CAT (Cusum Analysis Tool) to do these calculations can be downloaded from the web site www.niwa.cri.nz/.

Example 5.3 CUSUM Analysis of pH Data

As an example of the use of the CUSUM procedure, consider again the data from a Norwegian research programme that are shown in Table 5.3. Figure 5.9 shows the plots for the years 1976, 1977, 1978 and 1981, each compared to the mean for all years. It appears that the pH values in 1976 were generally close to the mean, with the CUSUM plots falling well within the limits found from 9,999 randomizations, and the Z_{max} value (1.76) being quite unexceptional ($p = 0.48$). For 1977 there is some suggestion of pH values being low for lakes with a low mean but the Z_{max} value (2.46) is not particularly large ($p = 0.078$). For 1978 there is also little evidence of differences from the mean for all years ($Z_{max} = 1.60$; $p = 0.626$). However, for 1981 it is quite another matter. The CUSUM plot even exceeds some of the upper limits obtained from 9,999 randomizations, and the observed Z_{max} value (3.96) is most unlikely to have been obtained by chance ($p = 0.0002$). Thus there is very clear evidence that the pH values in 1981 were higher than in the other years, although the CUSUM plot shows that the differences were largely on the lakes with relatively low mean pH levels because it becomes quite flat on the right-hand side of the graph.

The sum of the Z_{max} values for all four years is $Z_{max,T} = 9.79$. From the randomizations, the probability of obtaining a value this large is 0.0065, giving clear evidence of systematic changes in the pH levels of the lakes, for the years considered together.

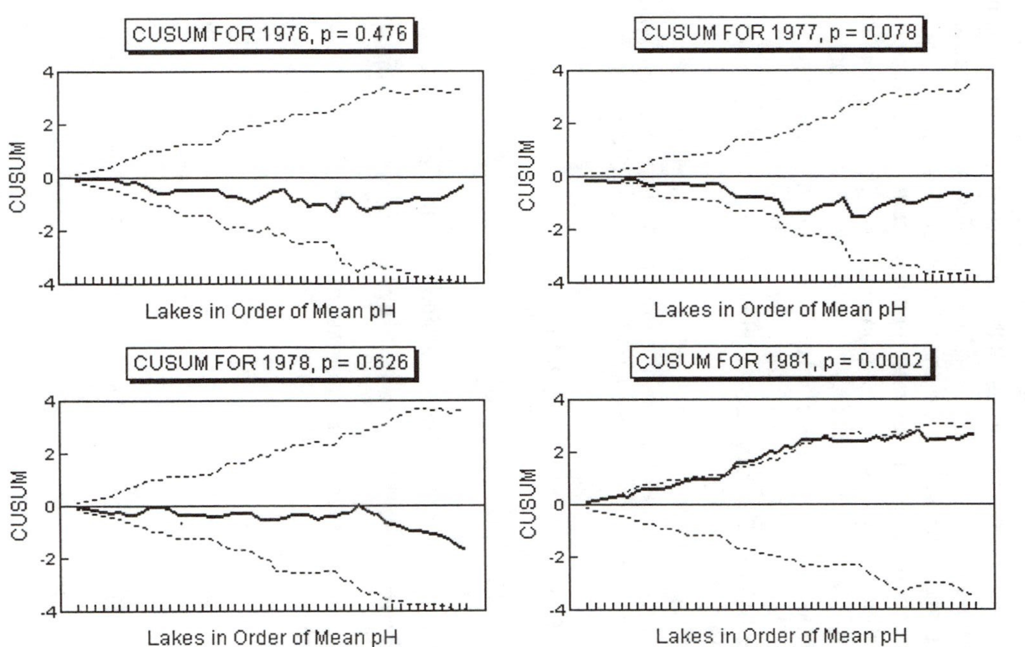

Figure 5.9 CUSUM plots for 1976, 1977, 1978 and 1981 compared to the mean for all years. The horizontal axis corresponds to lakes in the order of mean pH values for the four years. For each plot the upper and lower continuous lines are the limits obtained from 9,999 randomized sets of data and the CUSUM is the line with open boxes. The p-value with each plot is the probability of obtaining a plot as extreme or more extreme than the one obtained.

5.9 Chi-Squared Tests for a Change in a Distribution

Stehman and Overton (1994) describe a set of tests for a change in a distribution, which they suggest will be useful as a screening device. These tests can be used whenever observations are available on a random sample of units at two times. If the first observation in a pair is x and the second one is y, then y is plotted against x and three chi-squared calculations are made, as shown on Figure 5.10.

The first test compares the number of points above a 45 degree line with the number below, as indicated in Figure 5.10. A significant difference indicates an overall shift in the distribution either upward (most points above the 45 degree line) or downward (most points below the 45 degree line). In Figure 5.10 there are 30 points above the line and 10 points below. The expected counts are both 20 if x and y are from the same distribution.

The one sample chi-squared test (Appendix Section A4) is used to compare the observed and expected frequencies. The test statistic is $\Sigma(O - E)^2/E$, where O is an observed frequency, E an expected frequency, and the summation is over all such frequencies. Thus for the situation being considered, the statistic is

$$(30 - 20)^2/20 + (10 - 20)^2/20 = 10.00,$$

with 1 degree of freedom (df). This is significantly large at the 1% level, giving clear evidence of a shift in the general level of observations. Because most plotted points are above the 45 degree line the observations tend to be higher at the second sample time than at the first sample time.

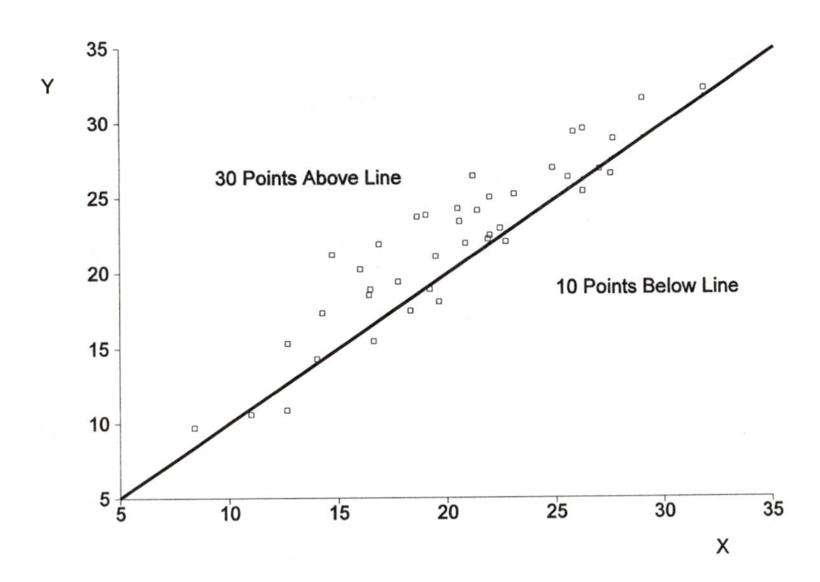

Figure 5.10 Test for a shift in a distribution by counting points above and below the line y = x.

For the second test, the line at 45 degrees is shifted upward or downward so that an equal number of points are above and below it. Counts are then made of the number of points in four quadrats, as shown in Figure 5.11. The counts then form a 2x2 contingency table, and the contingency table chi-squared test (Appendix Section A4) can be used to see whether the counts above and below the line are significantly different. A significant result indicates a change in shape of the distribution from one time period to the next. In Figure 5.11 the observed counts are

10 10
10 10.

These are exactly equal to the expected counts on the assumption that the probability of a point plotting above the line is the same for high and low observations, leading to a chi-squared value of zero with 1 df. There is therefore no evidence of a change in shape of the distribution from this test.

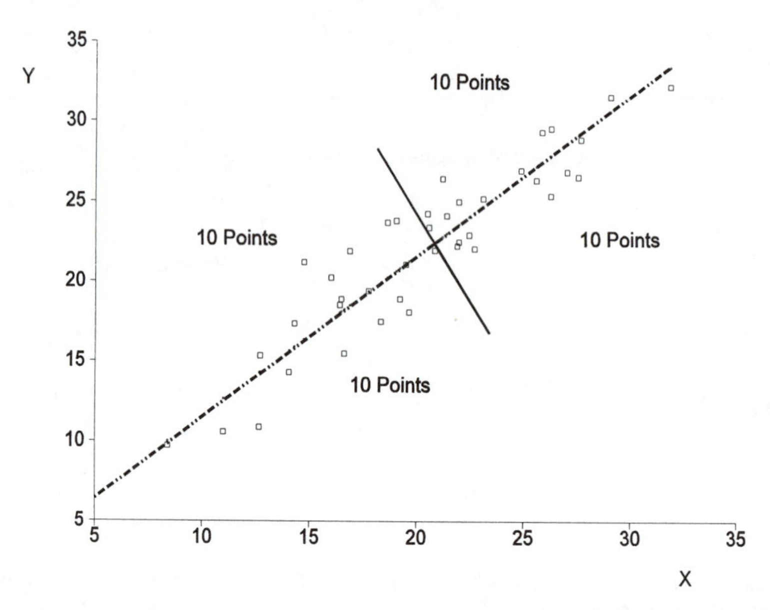

Figure 5.11 Test for a change of shape in a distribution by counting points around a shifted line.

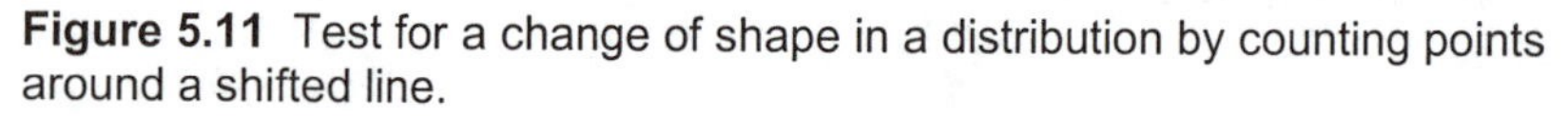

Finally, the third test involves dividing the points into quartiles, as shown in Figure 5.12. The counts in the different parts of the plot now make a 4x2 contingency table for which a significant chi-squared statistic again indicates a change in distribution. The contingency table from Figure 6 (c) is shown in Table 5.7. The chi-squared statistic is 0.80 with 3 df, again giving no evidence of a change in the shape of the distribution.

Example 5.4 The pH for Norwegian Lakes in 1976 and 1977

Figure 5.13 shows the 1978 pH values plotted against the 1976 values for the 44 lakes shown in Table 5.3 for which observations are available for both years. For the first of Stehman and Overton's (1994) tests, it is found that there are 13 points that plot above, and 31 that plot below the 45 degree line. The chi-square statistic is therefore

$$(13 - 22)^2/22 + (31 - 22)^2/22 = 7.36,$$

with 1 df. Because this is significant at the 1% level, there is clear evidence of a shift (downward) in the distribution.

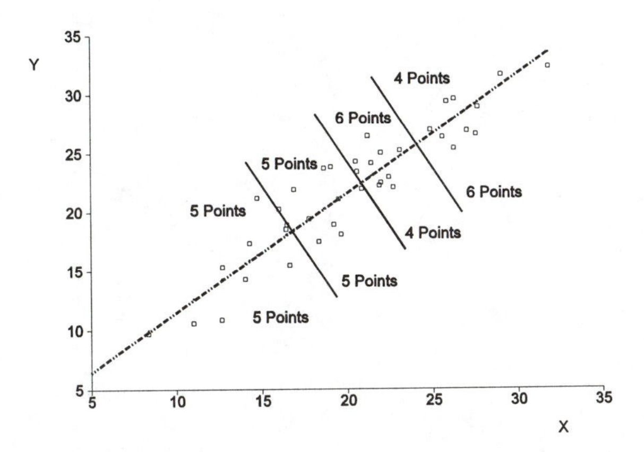

Figure 5.12 Extension of the test for a change in the shape of a distribution with more division of points.

Table 5.7 Counts of points above and below the shifted 45 degree line shown in Figure 5.10 (a)

	Quartile of the Distribution				
	1	2	3	4	Total
Above line	5	5	6	4	20
Below line	5	5	4	6	20
	10	10	10	10	40

For the second of Stehman and Overton's (1994) tests the counts above and below a shifted line are found to be

15 7
7 15.

The chi-squared statistic for this 2 x 2 contingency table is 5.82 with 1 df. This is significant at the 5% level, giving some evidence of a change in the shape of the distribution. The nature of the changes is that there seems to have been a tendency for low values to increase and high values to decrease between 1976 and 1978.

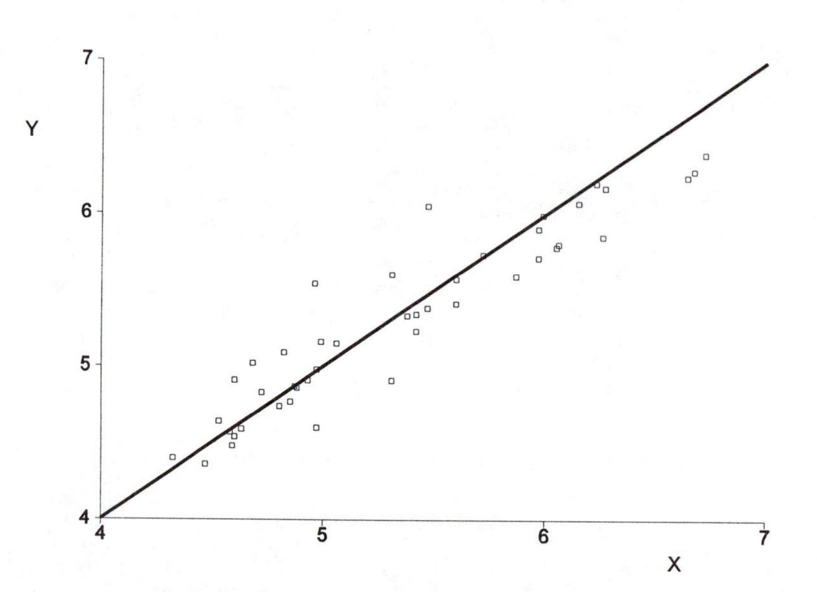

Figure 5.13 Plot of 1978 pH values (Y) against 1976 pH values (X) for 44 Norwegian lakes, for use with Stehman and Overton's tests.

For the third test, the counts are given in Table 5.8. The expected count in each cell is 5.5, and the contingency table chi-squared test comparing observed and expected frequencies gives a test statistic of 7.64 with 3 df. This is not quite significantly large at the 5% level, so that at this level of detail the evidence for a change in distribution is not strong.

Overall, the three chi-squared tests suggest that there has been a shift in the distribution towards lower pH values, with some reduction in variation as well.

Table 5.8 Counts of points above and below the shifted 45 degree line for the comparison between pH levels of lakes in 1976 and 1978

	Quartile of the Distribution				
	1	2	3	4	Total
Above line	7	8	5	2	22
Below line	4	3	6	9	22
	11	11	11	11	44

5.10 Chapter Summary

- A number of monitoring schemes are set up around the world to detect changes and trends in environmental variables, at scales ranging from whole countries to small local areas. Monitoring sites may be purposely chosen to be representative and meet other criteria, be chosen using special sampling designs incorporating both fixed sites and changing sites, or based on some optimisation principle.

- If repeated measurements are taken at a set of sites over a region, then one approach to testing for time changes is to use two factor analysis of variance. The first factor is then the site and the second factor is the sample time. This type of analysis is illustrated using pH data from Norwegian lakes.

- Control charts are widely used to monitor industrial processes, and can be used equally well to monitor environmental variables. With Shewhart's method there is one chart to monitor the process mean and a second chart to monitor the variability. The construction of these charts is described, and illustrated using pH data from rivers in the South Island of New Zealand.

- Cumulative sum (CUSUM) charts are an alternative to the Shewhart control chart for the mean. They involve plotting accumulating sums of deviations from a target value for a mean, rather than sample means. This type of chart may be able to detect small changes in the mean more effectively than the Shewhart chart, but is more complicated to set up.

- A variation on the usual CUSUM idea can be used to detect systematic changes with time in situations where a number of sampling stations are sampled at several points in time. This procedure incorporates randomization tests to detect changes, and is illustrated using the pH data from Norwegian lakes.

- A method for detecting changes in a distribution using chi-squared tests is described. With this test, the same randomly selected sample units are measured at two times, and the second measurement (y) is plotted against the first measurement (x). Based on the positions of the plotted points, it is possible to test for a change in the mean of the distribution, followed by tests for changes in the shape of the distribution.

CHAPTER 6

Impact Assessment

6.1 Introduction

The before-after-control-impact (BACI) sampling design is often used for assessing the effects of an environmental change made at a known point in time, and was called the 'optimal impact study design' by Green (1979). The basic idea is that one or more potentially impacted sites are sampled both before and after the time of the impact, and one or more control sites that cannot receive any impact are sampled at the same time. The assumption is that any naturally occurring changes will be about the same at the two types of sites, so that any extreme changes at the potentially impacted sites can be attributed to the impact. An example of this type of study is given in Example 1.4, where the chlorophyll concentration and other variables were measured on two lakes on a number of occasions from June 1984 to August 1986, with one of the lakes receiving a large experimental manipulation in the piscivore and planktivore composition in May 1985.

Figure 6.1 illustrates a situation where there are three observation times before the impact, and four observation times after the impact. Evidence for an impact is provided by a statistically significant change in the difference between the control and impact sites before and after the impact time. On the other hand, if the time plots for the two types of sites remain approximately parallel, then there is no evidence that the impact had an effect. Confidence in the existence of a lasting effect is also gained if the time plots are approximately parallel before the impact time, and then approximately parallel after the impact time, but with the difference between them either increased or decreased.

It is possible, of course for an impact to have an effect that increases or decreases with time. Figure 6.2 illustrates the latter situation, where the impacted site apparently returns to its usual state by about two time periods after the impact.

As emphasised by Underwood (1994), it is desirable to have more than one control site to compare with the potentially impacted site, and where possible these should be randomly selected from a population of sites that are physically similar to the impact site. It is also important to compare control sites to each other in order to be able to claim that the changes in the control sites reflect the changes that would be present in the impact site if there were no effect of the impact.

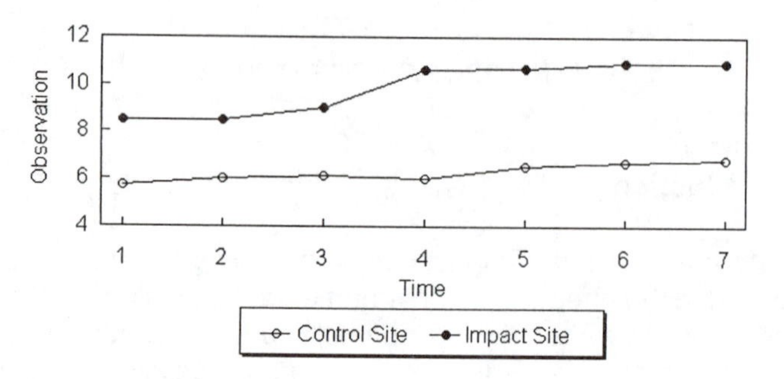

Figure 6.1 A BACI study with three samples before and four samples after the impact, which occurs between times 3 and 4.

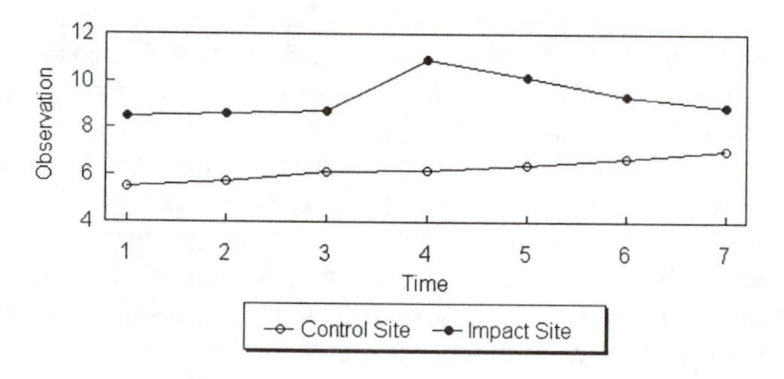

Figure 6.2 A situation where the effect of an impact between times 3 and 4 becomes negligible after 4 time periods.

In experimental situations, there may be several impact sites as well as several control sites. Clearly, the evidence of an impact from some treatment is much improved if about the same effect is observed when the treatment is applied independently in several different locations.

The analysis of BACI and other types of studies to assess the impact of an event may be quite complicated because there are usually repeated measurements taken over time at one or more sites. The repeated measurements at one site will then often be correlated, with those that are close in time tending to be more similar than those that are further apart in time. If this correlation exists but is not taken into account in the analysis of data, then the design has pseudoreplication

(Section 4.8), with the likely result being that the estimated effects appear to be more statistically significant than they should be.

When there are several control sites and several impact sites, each measured several times before and several times after the time of the impact, then one possibility is to use a repeated measures analysis of variance. The form of the data would be as shown in Table 6.1, which is for the case of three control and three impact sites, three samples before the impact, and four samples after the impact. For a repeated measures analysis of variance the two groups of sites give a single treatment factor at two levels (control and impact), and one within site factor which is the time relative to the impact, again at two levels (before and after). The repeated measurements are the observations at different times within the levels before and after, for one site. Interest is in the interaction between treatment factor and the time relative to the impact factor, because an impact will change the observations at the impact sites but not the control sites.

Table 6.1 The form of results from a BACI experiment with three observations before and four observations after the impact time. Observations are indicated by X

	Before the Impact			After the Impact			
Site	Time 1	Time 2	Time 3	Time 4	Time 5	Time 6	Time 7
Control 1	X	X	X	X	X	X	X
Control 2	X	X	X	X	X	X	X
Control 3	X	X	X	X	X	X	X
Impact 1	X	X	X	X	X	X	X
Impact 2	X	X	X	X	X	X	X
Impact 3	X	X	X	X	X	X	X

There are other analyses that can be carried out on data of the form shown in Table 6.1 that make different assumptions, and may be more appropriate, depending on the circumstances (Von Ende, 1993). Sometimes the data are obviously not normally distributed, or for some other reason a generalized linear model approach as discussed in Section 3.6 is needed rather than an analysis of variance. This is likely to be the case, for example, if the observations are counts or proportions. There are so many possibilities that it is not possible to cover them all here, and expert advice should be sought unless the appropriate analysis is very clear.

Various analyses have been proposed for the situation where there is only one control and one impact site (Manly, 1992, Chapter 6; Rasmussen *et al.*, 1993). In the next section a relatively straightforward approach is described that may properly allow for serial correlation in the observations from one site.

6.2 The Simple Difference Analysis with BACI Designs

Hurlebert (1984) highlighted the potential problem of pseudoreplication with BACI designs due to the use of repeated observations from sites. To overcome this, Stewart-Oaten *et al.* (1986) suggested that if observations are taken at the same times at the control and impact sites, then the differences between the impact and control sites at different times may be effectively independent. For example, if the control and impact sites are all in the same general area, then it can be expected that they will be affected similarly by rainfall and other general environmental factors. The hope is that by considering the difference between the impact and control sites the effects of these general environmental factors will cancel out.

This approach was briefly described in Example 1.4 on a large-scale perturbation experiment. The following is another example of the same type. Both of these examples involve only one impact site and one control site. With multiple sites of each type the analysis can be applied using the differences between the average for the impact sites and the average for the control sites at different times.

Carpenter *et al.* (1989) considered the question of how much the simple difference method is upset by serial correlation in the observations from a site. As a result of a simulation study, they suggested that to be conservative (in the sense of not declaring effects to be significant more often than expected by chance) results that are significant at a level of between 1% and 5% should be considered to be equivocal. This was for a randomization test, but their conclusion is likely to apply equally well to other types of tests such as the t-test used with Example 6.1.

Example 6.1 The Effect of Poison Pellets on Invertebrates

Possums (*Trichosurus vulpecula*) cause extensive damage in New Zealand forests when their density gets high, and to reduce the damage aerial drops of poison pellets containing 1080 (sodium monofluoroacetate) poison are often made. The assumption is made that the aerial drops have a negligible effect on non-target species, and a number of experiments have been carried out by the New Zealand Department of Conservation to verify this.

One such experiment was carried out in 1997, with one control and one impact site (McQueen and Lloyd, 2000). At the control site 100 non-toxic baits were put out on six occasions and the proportion of these that were fed on by invertebrates was recorded for three nights. At the impact site observations were taken in the same way on the same six occasions, but for the last two occasions the baits were toxic, containing 1080 poison. In addition, there was an aerial drop of poison pellets in the impact area between the fourth and fifth sample times. The question of interest was whether the proportion of baits being fed on by invertebrates dropped in the impact area after the aerial drop. If so, this may be result of the invertebrates being adversely affected by the poison pellets.

The available data are shown in Table 6.2, and plotted on Figure 6.3. The mean difference (impact - control) for times 1 to 4 before the aerial drop is -0.138. The mean difference after the drop for times 5 and 6 is -0.150, which is very similar. Figure 6.3 also shows that the time changes were rather similar at both sites, so there seems little suggestion of an impact. Treating the impact - control differences before the impact as a random sample of size 4, and the differences after the impact as a random sample of size 2, the change in the mean difference -0.150 - (-0.138) = -0.012 can be tested for significance using a two-sample t-test. This gives t = -0.158 with 4 df, which is not at all significant (p = 0.88 on a two-sided test). The conclusion must therefore be that there is no evidence here of an impact resulting from the aerial drop and the use of poison pellets.

If a significant difference had been obtained from this analysis it would, of course, be necessary to consider the question of whether this was just due to the time changes at the two sites being different for reasons completely unrelated to the use of poison pellets at the impact site. Thus the evidence for an impact would come down to a matter of judgement in the end.

Table 6.2 Results from an experiment to assess whether the proportion of pellets fed on by invertebrates changes when the pellets contain 1080 poison.

Time	Control	Impact	Difference
1	0.40	0.37	-0.03
2	0.37	0.14	-0.23
3	0.56	0.40	-0.16
4	0.63	0.50	-0.13
	Start of Impact		
5	0.33	0.26	-0.07
6	0.45	0.22	-0.23
		Mean Difference	
		Before	-0.138
		After	-0.150

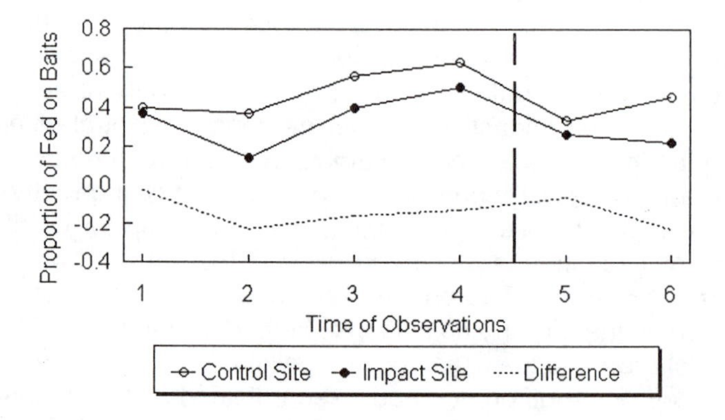

Figure 6.3 Results from a BACI experiment to see whether the proportion of pellets fed on by invertebrates changes when there is an aerial drop of 1080 pellets at the impact site between times 4 and 5.

6.3 Matched Pairs with a BACI Design

When there is more than one impact site, pairing is sometimes used to improve the study design, with each impact site being matched with a control site that is as similar as possible. This is then called a control-treatment paired (CTP) design (Skalski and Robson, 1992, Chapter 6), or a before-after-control-impact-pairs (BACIP) design (Stewart-Oaten *et al.*, 1986). Sometimes information is also collected on variables that describe the characteristics of the individual sites (elevation, slope, etc.). These can then be used in the analysis of the data to allow for imperfect matching. The actual analysis depends on the procedure used to select and match sites, and on whether or not variables to describe the sites are recorded.

The use of matching can lead to a relatively straightforward analysis, as demonstrated by the following example.

Example 6.2 Another Study of the Effect of Poison Pellets

Like Example 6.1, this concerns the effects of 1080 poison pellets on invertebrates. However, the study design was rather different. The original study is described by Sherley and Wakelin (1999). In brief, 13 separate trials of the use of 1080 were carried out, where for each trial about 60 pellets were put out in a grid pattern in two adjacent sites, over each of nine successive days. The pellets were of the type used in aerial drops to reduce possum numbers. However, in one of the two adjacent sites used for each trial the pellets never contained 1080 poison. This served as the control. In the other site the pellets contained poison on days 4, 5 and 6 only. Hence the control and impact sites were observed for three days before the impact, for three days during the impact (1080 pellets), and for three days after the impact was removed. The study carried out by Sherley and Wakelin involved some other components as well as the nine day trials, but these will not be considered here.

The average number of invertebrates seen on pellets each day is shown in the top graph of Figure 6.4, for each of the 13 x 2 = 26 sites. There is a great deal of variation in these averages, although it is noticeable that the control sites tend to higher means, as well as being more variable than the poison sites. When the results are averaged for the control and poison sites, a clearer picture emerges (Figure 6.4, bottom graph). The poison sites had slightly lower mean counts than the control sites for days 1 to 3, the mean for the poison sites was much lower for days 4 to 6, and then the difference became less for days 7 to 9.

If the differences between the pairs of sites are considered, then the situation becomes somewhat clearer (Figure 6.5). The poison sites always had a lower mean than the control sites, but the difference increased for days 4 to 6, and then started to return to the original level.

Once differences are taken, a result is available for each of the nine days, for each of the 13 trials. An analysis of these differences is possible using a two factor analysis of variance, as discussed in Section 3.5. The two factors are the trial at 13 levels, and the day at nine levels. As there is only one observation for each combination of these levels, it is not possible to estimate an interaction term, and the model

$$x_{ij} = \mu + a_i + b_j + \epsilon_{ij} \qquad (6.1)$$

must be assumed, where x_{ij} is the difference for trial i on day j, μ is an overall mean, a_i is an effect for the ith trial, b_j is an effect for the jth day, and ϵ_{ij} represents random variation. When this model was fitted using Minitab (Minitab Inc., 1994) the differences between trials were highly significant ($F = 8.89$ with 12 and 96 df, $p < 0.0005$), as were the differences between days ($F = 9.26$ with 8 and 96 df, $p < 0.0005$). It appears, therefore, that there is very strong evidence that the poison and control sites changed during the study, presumably because of the impact of the 1080 poison.

There may be some concern that this analysis will be upset by serial correlation in the results for the individual trials. However, this does not seem to be a problem here because there are wide fluctuations from day to day for some trials (Figure 6.5). Of more concern is the fact that the standardized residuals (the differences between the observed values for x and those predicted by the fitted model, divided by the estimated standard deviation of the error term in the model) are more variable for the larger predicted values (Figure 6.6). This seems to be because the original counts of invertebrates on the pellets have a variance that increases with the mean value of the count. This is not unexpected because it is what usually occurs with counts, and a more suitable analysis for the data involves fitting a log-linear model (Section 3.6) rather than an analysis of variance model. However, if a log-linear model is fitted to the count data, then exactly the same conclusion is reached: the difference between the poison and control sites changes systematically over the nine days of the trials, with the number of invertebrates decreasing during the three days of poisoning at the treated sites, followed by some recovery towards the initial level in the next three days.

This conclusion is quite convincing because of the replicated trials and the fact that the observed impact has the pattern that is expected if the 1080 poison has an effect on invertebrate numbers. The same conclusion was reached by Sherley and Wakelin (1999) but using a randomization test instead of analysis of variance or log-linear modelling.

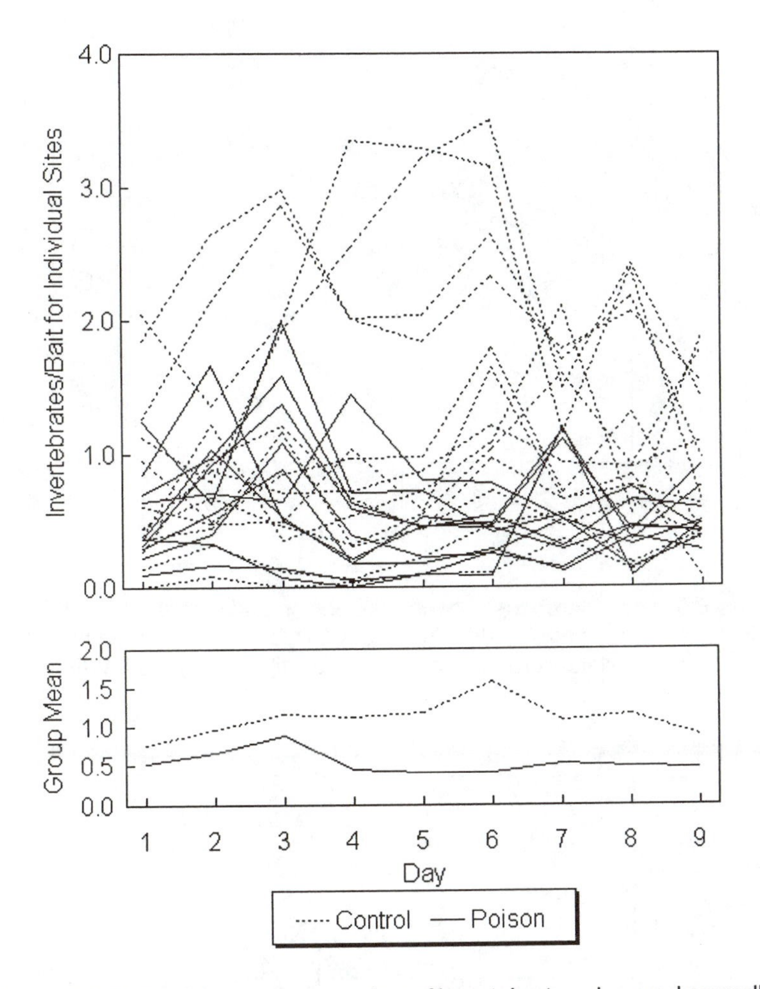

Figure 6.4 Plots of the average number of invertebrates observed per pellet (top graph) and the daily means (bottom graph) for the control areas (broken lines) and the treated areas (continuous lines). At the treated site poison pellets were used on days 4, 5 and 6 only.

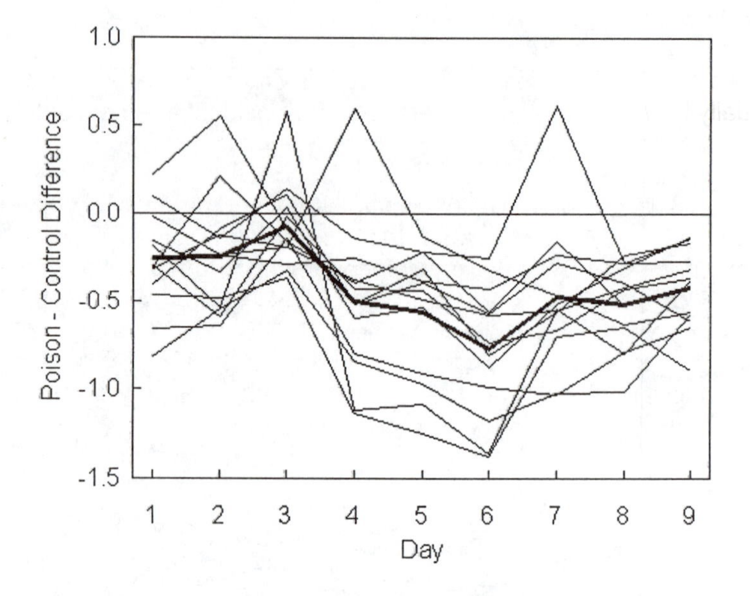

Figure 6.5 The differences between the poison and control sites for the 13 trials, for each day of the trials. The heavy line is the mean difference for all trials. Poison pellets were used at the treated site for days 4, 5 and 6.

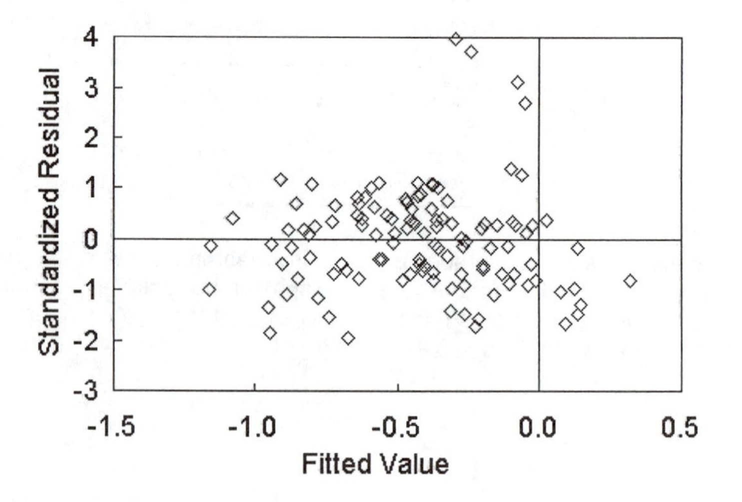

Figure 6.6 Plot of the standardized residuals from a two factor analysis of variance against the values predicted by the model, for the difference between the poison and control sites for one day of one trial.

6.4 Impact-Control Designs

When there is an unexpected incident such as an oil spill there will usually be no observations taken before the incident at either control or impact sites. The best hope for impact assessment then is the impact-control design, which involves comparing one or more potentially impacted sites with similar control sites. The lack of 'before' observations typically means that the design has low power in comparison with BACI designs (Osenberg *et al.*, 1994).

It is obvious that systematic differences between the control and impact sites following the incident may be due to differences between the types of sites rather than the incident. For this reason, it is desirable to measure variables to describe the sites, in the hope that these will account for much of the observed variation in the variables that are used to describe the impact, if any.

Evidence of a significant area by time interaction is important in an impact-control design, because this may be the only source of information about the magnitude of an impact. For example, Figure 6.7 illustrates a situation where there is a large immediate effect of an impact, followed by an apparent recovery to the situation where the control and impact areas become rather similar.

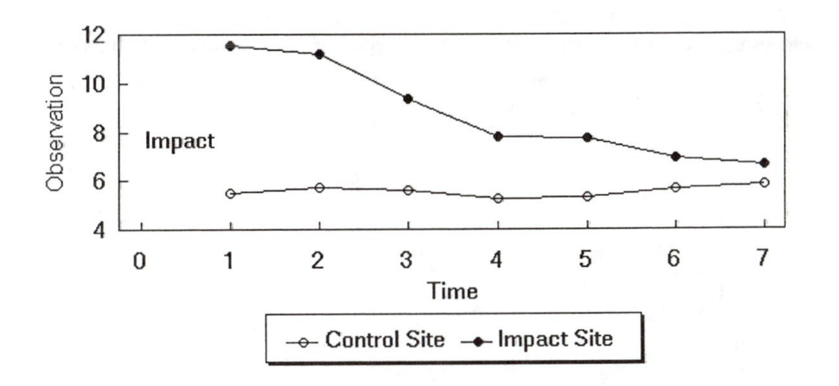

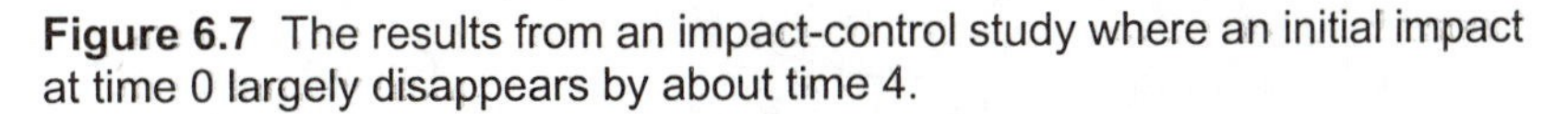

Figure 6.7 The results from an impact-control study where an initial impact at time 0 largely disappears by about time 4.

The analysis of the data from an impact-control study will obviously depend on precisely how the data are collected. If there are a number of control sites and a number of impact sites measured once each, then the means for the two groups can be compared by a standard test of significance, and confidence limits for the difference can be calculated. If each site is measured several times, then a repeated measures analysis of variance may be appropriate. The sites are then the 'subjects', with the two groups of sites giving two levels of a treatment factor. As with the BACI design with multiple sites, careful thought is needed to choose the best analysis for these types of data.

6.5 Before-After Designs

The before-after design can be used for situations where either no suitable control areas exist, or it is not possible to measure suitable areas. It does requires data to be collected before a potential impact occurs, which may be the case with areas that are known to be susceptible to damage, or which are being used for long-term monitoring. The key question is whether the observations taken immediately after an incident occurs can be considered to fit within the normal range for the system. A pattern such as that shown in Figure 6.8 is expected, with a large change after the impact followed by a return to normal conditions.

The analysis of the data must be based on some type of time series analysis as discussed in Chapter 8 (Manly, 1992, Chapter 6; Rasmussen *et al.*, 1993). In simple cases where serial correlation in the observations is negligible a multiple regression model may suffice. However, if serial correlation is clearly present, then this should be allowed for, possibly using a regression model with correlated errors (Neter *et al.*, 1983, Chapter 13).

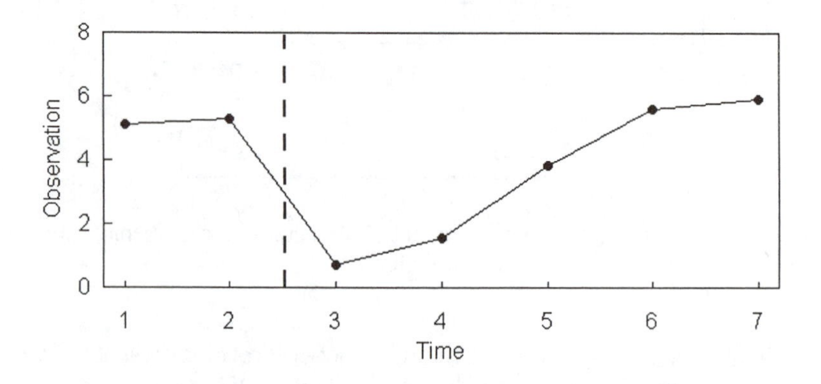

Figure 6.8 The before-after design where an impact between times 2 and 3 disappears by about time 6.

Of course, if some significant change is observed it is important to be able to rule out causes other than the incident. For example, if an oil spill occurs because of unusually bad weather, then the weather itself may account for large changes in some environmental variables, but not others.

6.6 Impact-Gradient Designs

The impact-gradient design can be used where there is a point source of an impact, in areas that are fairly homogeneous. The idea is to establish a function which demonstrates that the impact reduces as the distance from the source of the impact increases. To this end, data are collected at a range of distances from the source of the impact, preferably with the largest distances being such that no impact is expected. Regression methods can then be used to estimate the average impact as a function of the distance from the source. There may well be natural variation over the study area associated with the type of habitat at different sample locations, in which case suitable variables should be measured so that these can be included in the regression equation to account for the natural variation as far as possible.

A number of complications can occur with the analysis of data from the impact-gradient design. The relationship between the impact and the distance from the source may not be simple, necessitating the use of non-linear regression methods, the variation in observations may not be constant at different distances from the source, and there may be spatial correlation, as discussed in Chapter 9. This is therefore another situation where expert advice on the data analysis may be required.

6.7 Inferences from Impact Assessment Studies

'True' experiments as defined in Section 4.3 include randomization of experimental units to treatments, replication to obtain observations under the same conditions, and control observations that are obtained under the same conditions as observations with some treatment applied, but without any treatment. Most studies to assess environmental impacts do not meet these conditions, and hence result in conclusions that must be accepted with reservations. This does not mean that the conclusions are wrong. It does mean that alternative explanations for observed effects must be ruled out as unlikely to be true.

It is not difficult to devise alternative explanations for the simpler study designs. With the impact-control design (Section 6.4) it is always possible that the differences between the control and impact sites existed before the time of the potential impact. If a significant difference is observed after the time of the potential impact, and this is claimed to be a true measure of the impact, then this can only be based on the judgement that the difference is too large to be part of normal variation. Likewise, with the before-after design (Section 6.5), if the change from before to after the time of the potential impact is significant and this is claimed to represent the true impact, then this is again based on a judgement that the magnitude of the change is too large to be explained by anything else. Furthermore, with these two designs no amount of complicated statistical analysis can change these basic weaknesses. In the social science literature these designs are described as pre-experimental designs because they are not even as good as quasi-experimental designs.

The BACI design with replication of control sites at least is better because there are control observations in time (taken before the potential impact) and in space (the sites with no potential impact). However, the fact is that just because the control and impact sites have an approximately constant difference before the time of the potential impact it does not mean that this would necessarily continue in the absence of an impact. If a significant change in the difference is used as evidence of an impact, then it is an assumption that nothing else could cause a change of this size.

Even the impact-gradient study design is not without its problems. It might seem that a statistically significant trend in an environmental variable with increasing distance from a potential point source of an impact is clear evidence that the point source is responsible for the change. However, the variable might naturally display spatial patterns and trends associated with obvious and non-obvious physical characteristics of the region. The probability of detecting a spurious significant trend may then be reasonably high if this comes from an analysis that does not take into account spatial correlation.

With all these limitations, it is possible to wonder about the value of many studies to assess impacts. The fact is that they are often done because they are all that can be done, and they give more information than no study at all. Sometimes the estimated impact is so large that it is impossible to imagine it being the result of anything but the impact event, although some small part of the estimate may indeed be due to natural causes. At other times the estimated impact is small and insignificant, in which case it is not really possible to argue that somehow the real impact is really large and important.

6.8 Chapter Summary

- The before-after-control-impact (BACI) study design is often used to assess the impact of some event on variables that measure the state of the environment. The design involves repeated measurements over time being made at one or more control sites and one or more potentially impacted sites, both before and after the time of the event that may cause an impact.

- Serial correlation in the measurements taken at a site results in pseudoreplication if it is ignored in the analysis of data. Analyses that may allow for this serial correlation in an appropriate way include repeated measures analysis of variance.

- A simple method that is valid with some sets of data takes the differences between the observations at an impact site and a control site, and tests for a significant change in the mean difference from before the time of the potential impact to after this time. This method can be applied using the differences between the mean for several impact sites and the mean for several control sites. It is illustrated using the results of an experiment on the effect of poison pellets on invertebrate numbers.

- A variation of the BACI design uses control and impact sites that are paired up on the basis of their similarity. This can allow a relatively simple analysis of the study results, as is illustrated by another study on the effect of poison pellets on invertebrate numbers.

- With an impact-control design, measurements at one or more control sites are compared with measurements at one or more impact sites only after the potential impact event has occurred.

- With a before-after design, measurements are compared before and after the time of the potential impact event, at impact sites only.

- An impact-gradient study can be used when there is a point source of a potential impact. This type of study looks for a trend in the values of an environmental variable with increasing distance from the point source.

- Impact studies are not usually true experiments with randomization, replication and controls. The conclusions drawn are therefore based on assumptions and judgement. Nevertheless, they are often carried out because nothing else can be done and they are better than no study at all.

CHAPTER 7

Assessing Site Reclamation

7.1 Introduction

This chapter is concerned with the specific problem of evaluating the effectiveness of the reclamation of a site that has suffered from some environmental damage. An example of the type of situation to be considered is where a site has been used for mining in the past and a government agency now requires that the mining company improve the state of the site until the biomass of vegetation per unit area is similar to what is found on an undamaged reference site.

There are some difficulties with treating this problem using a classical test of significance. These are discussed in the next section of the chapter. An alternative approach that has gained support from some environmental scientists and managers is to use the concept of bioequivalence for comparing the sites. Much of the chapter is concerned with how this alternative approach can be applied.

7.2 Problems with Tests of Significance

At first sight it might seem that it is a straightforward problem to decide whether two sites are similar in terms of something like the biomass of vegetation, and that this can be dealt with in the following manner. The damaged site should be improved until it appears to be similar to the reference site. Random sample quadrats should then be taken from each of the sites and the mean biomass calculated. If the two means are not significantly different, then the two sites are declared to be 'similar'.

Unfortunately, as noted in Example 1.7 which was concerned with this type of problem, there are two complications with this obvious approach:

- It is unreasonable to suppose that the damaged and reference sites would have had exactly the same mean for the study variable, even in the absence of any impact on the damaged site. Therefore, if large samples are taken from each site, there will be a high probability of detecting a difference, irrespective of the extent to which the damaged site has been reclaimed. Hence, the question of interest should not be whether there is a significant difference between the sites. Rather, the question should be whether the difference is of practical importance.

- When a test for a difference between the two sites does not give a significant result, this does not necessarily mean that a difference does not exist. An alternative explanation is that the sample sizes were not large enough to detect the difference which does exist.

Given this situation, the mining company has two sensible options. It can try to ensure that the comparison of sites is done with the smallest possible sample sizes so that there is not much power to detect a small difference between the sites. Or alternatively, it can improve the damage site so that the biomass is much higher than for the reference site, on the assumption that the government agency will think this is acceptable. Neither of these options seems very satisfactory.

To avoid these complications with statistical tests, the United States Environmental Protection Agency (1989a) recommends that the null hypothesis for statistical tests should depend on the status of a site, in the following way:

(a) If a site has not been declared to be damaged, then the null hypothesis should be that it is not, i.e., there is no difference from the control site. The alternative hypothesis is that the site is contaminated. A non-significant test result leads to the conclusion that there is no real evidence that the site is damaged.

(b) If a site has been declared to be damaged then the null hypothesis is that this is true, i.e., there is a difference (in an unacceptable direction) from the control site. The alternative hypothesis is that the site is undamaged. A non-significant test result leads to the conclusion that there is no real evidence that the site has been cleaned up.

The point here is that once a site has been declared to have a certain status pertinent evidence should be required to justify changing this status.

Following these recommendations does seem to overcome the main difficulty with using a test of significance, although there is still the problem of deciding what to use for the null hypothesis difference if option (b) is used.

7.3 The Concept of Bioequivalence

When the null hypothesis to be tested is that a site is damaged, there is a need to define what exactly 'damaged' means. The concept of bioequivalence then becomes useful (McBride *et al.*, 1993; McDonald and Erickson, 1994; McBride, 1999). In the pharmaceutical area a new drug is considered to be 'bioequivalent' to a standard drug if the potency of the new drug is (say) at least 80% of the potency of the standard drug (Kirkwood, 1981; Westlake, 1988). In a similar way, a damaged site might be considered to be bioequivalent to a control site in terms of vegetation biomass if the mean biomass per unit area on the damaged site, μ_t, is at least 80% of the mean on the control site, μ_c. In that case, bioequivalence can be examined by testing the null hypothesis

$$H_0: \mu_t \leq 0.8\mu_c$$

against the alternative hypothesis

$$H_1: \mu_t > 0.8\mu_c.$$

Example 7.1 Native Shrubs at Reclaimed and Reference Sites

As an example of how the concept of bioequivalence might be used to assess reclamation, consider the following hypothetical situation described by McDonald and Erickson (1994), noting that the analysis here is simpler than the one that they used. It is imagined that a mining company has paid a bond to a government agency to guarantee the successful reclamation of a strip mining site. Having carried out the necessary work, the company wants the bond released. However, the agency requires the company to provide evidence that the mined site is equivalent to an untouched control site with respect to the density of native shrubs.

A consultant has designed and carried out a study that involved randomly selecting eight plots from the treated site and matching them up on the basis of slope, aspect, and soil type with eight plots from the control site. The densities of native shrubs that were obtained are shown in Table 7.1. The control - mined site differences are also shown with their means and sample standard deviations.

A conventional approach for analysing these results involves using a t-test to see whether the mean difference of $\bar{d} = 0.041$ is significantly greater than zero. The null hypothesis is then that the mean density of native shrubs is the same on paired plots at the two sites, while the alternative hypothesis is that the density is higher on the control site. The test statistic is

$$t = \bar{d} / SE(\bar{d}),$$

where $SE(\bar{d}) = SD(d)/\sqrt{n} = 0.171/\sqrt{8} = 0.060$ is the estimated standard error of the mean. That is, $t = 0.041/0.060 = 0.68$, with seven degrees of freedom (df). This is not significantly large at the 5% level because the critical value that has to be exceeded to make this the case is 1.89. The mining company can therefore argue that the reclamation has been effective.

Table 7.1 Comparison between the vegetation density on eight paired plots from an undamaged control site and a site where mining has occurred. The difference is for the control - mined

Plot pair	1	2	3	4	5	6	7	8
Control site	0.94	1.02	0.80	0.89	0.88	0.76	0.71	0.75
Mined site	0.75	0.94	1.01	0.67	0.75	0.88	0.53	0.89
Difference	0.19	0.08	-0.21	0.22	0.13	-0.10	0.18	-0.14

Mean difference = 0.041, Standard deviation of difference = 0.171

The government agency could object to this analysis on the grounds that the non-significant result may just be a result of the small sample size. They might well prefer an analysis which is based on the idea that the control and mined site are 'equivalent' for all practical purposes providing that the native shrub density on the mined site is more than 80% of the density on the control site. On this basis the null hypothesis is that the native shrub density at the mined site is 80% of the density at the control site, and the contrast

$$z = (\text{mined site density}) - 0.8 \times (\text{control site density})$$

will have a mean of zero for paired sites. The alternative hypothesis is that the mean of z is greater than zero, in which case the two sites are considered to be equivalent.

Note that now the null hypothesis is that the sites are not equivalent. The data have to provide evidence that this is not true before the sites are declared to be equivalent. Thus the precautionary principle is used: an adverse effect is assumed unless the data suggest otherwise.

The test procedure follows the same steps as the first analysis except that values of z are used instead of the simple differences between the paired sites, as shown in Table 7.2. The mean of the z values is 0.127, with an estimated standard error of $0.163/\sqrt{8} = 0.058$. The t-statistic for testing whether the mean is significantly greater than zero is therefore $0.127/0.058 = 2.21$, with seven df. Because this is significantly large at the 5% level, it is concluded that there is evidence against the null hypothesis and the equivalence of the mined and control site can be accepted.

Table 7.2 Testing for bioequivalence using the vegetation density on eight paired plots from an undamaged control site and a site where mining has occurred. The z value is the mined site density - 0.8 times the control size density

Plot pair	1	2	3	4	5	6	7	8
Control site	0.94	1.02	0.80	0.89	0.88	0.76	0.71	0.75
Mined site	0.75	0.94	1.01	0.67	0.75	0.88	0.53	0.89
z value	0.00	0.12	0.37	-0.04	0.05	0.27	-0.04	0.29

Mean of z = 0.127, Standard deviation of z = 0.163

This second analysis seems more realistic than the first one because the acceptance of the null hypothesis, possibly because of the small sample size, will result in the mined site being considered to need further remediation: the mined site is 'guilty' until proved 'innocent', rather than 'innocent' until proved 'guilty'. The definition of equivalence in terms of the mined site having more than 80% of the shrub density of the control site would, of course, have been the subject of negotiations between the mining company and the government agency. Another percentage could be used equally well in the test.

7.4 Two-Sided Tests of Bioequivalence

The example just considered was quite straightforward because the test was one-sided, and the data were paired. A more complicated situation is where a previously damaged site is considered to be equivalent to an undamaged reference site providing that the mean of a relevant variable at the first site is sufficiently close to the mean at the reference site.

Here the null hypothesis can be that the two sites are not equivalent (following the precautionary principle) or that they are equivalent. In the first case the null hypothesis becomes that $\mu_d < \mu_{dL}$ or $\mu_d > \mu_{dH}$, where the two sites are considered to be equivalent if μ_d, the true difference between them (damaged - reference), is within the range from μ_{dL} to μ_{dH}. In the second case the null hypothesis is that $\mu_{dL} \leq \mu_d \leq \mu_{dH}$. It may be very important which of these null hypotheses is chosen because with the first a significant result leads to the conclusion that the two sites are equivalent, whereas with the second a significant result leads to the conclusion that the sites are not equivalent.

The simplest way to test the null hypothesis that the two sites are not equivalent is to run the two one-sided test (TOST) developed by Schuirmann (1987) and Westlake (1988). Assuming normally distributed data, with equal variances for the potentially damaged site and the reference site, this proceeds as follows for a 5% level of significance:

(a) Calculate the mean difference $\bar{d}$ between the potentially damaged site and the reference site, and the estimated standard error of this difference

$$SE(\bar{d}) = s_p\sqrt{(1/n_1 + 1/n_2)}$$

where n_1 is the sample size for the damaged site and n_2 is the sample size for the reference site,

$$s_p^2 = \{(n_1 - 1)s_1^2 + (n_2 - 1)s_2^2\}/(n_1 + n_2 - 2)$$

is the pooled-sample estimate of variance, s_1^2 is the sample variance for the damaged site, and s_2^2 is the sample variance for the reference site.

(b) Use a t-test to see whether $\bar{d}$ is significantly higher than μ_{dL} at the 5% level, which involves seeing whether $(\bar{d} - \mu_{dL})/SE(\bar{d})$ is greater than or equal to the upper 5% point of the t-distribution with $n_1 + n_2 - 2$ df.

(c) Use a t-test to see whether $\bar{d}$ is significantly lower than μ_{dH} at the 5% level, which involves seeing whether $(\bar{d} - \mu_{dH})/SE(\bar{d})$ is less than or equal to the lower 5% point of the t-distribution with $n_1 + n_2 - 2$ df.

(d) If the tests at steps (b) and (c) are both significant, then declare that there is evidence for the equivalence of the two sites. The logic here is that if the observed difference is both significantly higher than the lowest allowed difference, and also significantly lower than the highest allowed difference, then there is certainly evidence that it is within the allowed range.

Of course, this test can be carried out using a different significance level if necessary, and it should be noted that although it includes two t-tests there is no need to allow for multiple testing because the probability of declaring the two sites to be equivalent when they are not is no more than α if the two t-tests are each carried out at the $100\alpha\%$ level (Berger and Hsu, 1996).

If the null hypothesis is that the sites are equivalent ($\mu_{dL} \leq \mu_d \leq \mu_{dH}$), then the two tests that are part of the TOST procedure must be modified. Part (b) of the above procedure changes to:

(b') Use a t-test to see whether $\bar{d}$ is significantly lower than μ_{dL} at the 5% level, which involves seeing whether $(\bar{d} - \mu_{dL})/SE(\bar{d})$ is less than or equal to the lower 5% point of the t-distribution with $n_1 + n_2 - 2$ df.

This is then seeing whether there is any evidence that the true mean difference is lower than μ_{dL}. Similarly, part (c) of the procedure changes to:

(c') Use a t-test to see whether $\bar{d}$ is significantly higher than μ_{dH} at the 5% level, which involves seeing whether $(\bar{d} - \mu_{dH})/SE(\bar{d})$ is greater than or equal to the upper 5% point of the t-distribution with $n_1 + n_2 - 2$ df.

Now, if either of these tests gives a significant result, then there is evidence that the two sites are not equivalent.

The test of the non-equivalence null hypothesis is more stringent than the test of the equivalence null hypothesis because evidence is required before sites are declared to be equivalent, rather than the other way round. With the non-equivalence null hypothesis the TOST procedure carried out with a 5% level of significance can be shown to give evidence of equivalence if the sample mean difference falls in the interval

$$\mu_{dL} + t_{0.05,v}\,SE(\bar{d}) \le \bar{d} \le \mu_{dH} - t_{0.05,v}\,SE(\bar{d}), \qquad (7.1)$$

where $t_{0.05,v}$ is the value that is exceeded with probability 0.05 for the t-distribution with $v = n_1 + n_2 - 2$ df. On the other hand, with the equivalence null hypothesis carried out with the same level of significance there is no evidence against the null hypothesis if

$$\mu_{dL} - t_{0.05,v}\,SE(\bar{d}) \le \bar{d} \le \mu_{dH} + t_{0.05,v}\,SE(\bar{d}). \qquad (7.2)$$

The second interval may be much wider than the first one. This is demonstrated in Figure 7.1 which is for a hypothetical situation where two sites are considered to be equivalent if the mean difference is between -1 and +1.

There are procedures other than TOST for carrying out two-sided tests of bioequivalence, as reviewed by McBride (1999). Apparently the general view in the pharmaceutical literature, where most applications have been in the past, is that the TOST approach is best.

In Example 7.1 bioequivalence was expressed in terms of a ratio, with the equivalence of a damaged and a reference site being defined as the biomass per unit area of native plants in the damaged site being at least 80% of the value for the reference site. The two-sided version for this might then be that two sites are considered as equivalent providing that the ratio R = (density of native plants in an impacted area)/(density of native plants in a control area) should be within the range 0.8 to 1.2. McDonald and Erickson (1994) discuss procedures for use with this ratio type of approach.

Specialized computer programs are now available to carry out bioequivalence tests. One is EquivTest from Statistical Solutions (web site: www.statsolusa.com), and another is Power and Sample Size Analysis (PASS) from Number Cruncher Statistical Systems (web site: www.ncss.com).

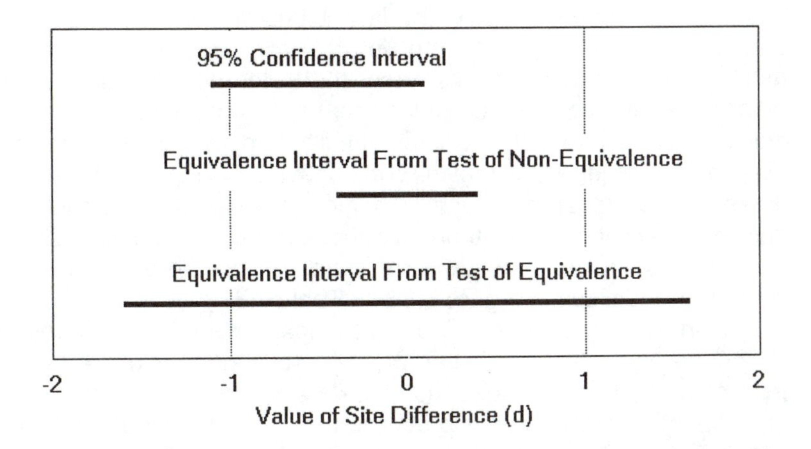

Figure 7.1 Bioequivalence intervals for a situation where two sites are considered to be equivalent if their true mean difference is between -1 and +1. It is assumed that a random sample of size 10 is taken from each of the two sites, and gives a sample mean difference of $\bar{d} = -0.5$ with an estimated standard error of $SE(\bar{d}) = 0.3$. The top interval is the 95% confidence interval for the true mean difference between the sites, $\bar{d} \pm 2.10\ SE(\bar{d})$, the middle interval is the range of sample means that give evidence for equivalence calculated from equation (7.1), and the bottom interval is the range of sample means that give no evidence against the hypothesis of equivalence calculated from equation (7.2).

Example 7.2 PCB at the Armagh Compressor Station

For an example of a comparison between a reference site and a potentially contaminated site, some data were extracted from a much larger set described by Gore and Patil (1994). Their study involved two phases of sampling of polychlorinated biphenyl (PCB) at the site of the Armagh compressor station in Indiana County, Pennsylvania, USA. The phase 1 sampling was in areas close to sources of PCB, while the phase 2 sampling was away from these areas. For the present purpose, a random sample of 30 observations was extracted from the phase 2 sampling results to represent a sample from a reference area, and a random sample of 20 observations was extracted from the phase 1 sample results to represent a sample from a possibly contaminated area.

The values for the PCB concentrations in parts per million (ppm) are shown in the left-hand side of Table 7.3, and plotted on the left-hand side of Figure 7.2. Clearly, the possibly contaminated sample has much more variable results than the reference sample, which complicates the comparison of the means. However, for data of this type it is common to find that distributions are approximately lognormal (Section 4.3), suggesting that the comparison between samples is best made on the logarithms of the original results, which should be approximately normally distributed with the variation being more similar in different samples. This turns out to be the case here, as shown by the right-hand sides of Figure 7.2 and Table 7.3.

It is in fact convenient to work with logarithms if it is desirable to define the equivalence between the two areas in terms of the ratio of their means. Thus suppose that it is decided that the two areas are equivalent in practical terms providing that the ratio of the mean PCB concentration in the possibly contaminated area to the mean in the reference area is between 0.5 and 1.0/0.5 = 2.0. Then this corresponds to a difference between the logarithms of mean of between log(0.5) = -0.301 and log(2.0) = +0.301, using logarithms to base 10. Then for the tests of non-equivalence and equivalence described above, μ_{dL} = -0.301, and μ_{dH} = +0.301. These tests will be carried out here using the 5% level of significance.

From the logarithmic data in Table 7.3, the observed mean difference between the samples is $\bar{d}$ = 0.630, with estimated standard error SE($\bar{d}$) = 0.297. For the test for non-equivalence, it is first necessary to see whether $\bar{d}$ is significantly higher than -0.301, at the 5% level of significance. The t-statistic is $t = (\bar{d} - \mu_{dL})/SE(\bar{d}) = 3.137$, with 48 df. The probability of a value this large or larger is 0.001, so there is evidence that the observed mean is higher than the lowest value allowed. Next, it is necessary to test whether $\bar{d}$ is significantly lower than +0.301, at the 5% level of significance. As $\bar{d}$ exceeds 0.301, this is clearly not true. This non-significant result means that the null hypothesis of non-equivalence is accepted. The conclusion is that there is no evidence that the areas are equivalent.

Table 7.3 PCB concentrations in a reference area and a possibly contaminated area around the Armagh compressor station, and results transformed to logarithms to base 10

	Original PCB Concentration (ppm)		After Log Transformation	
	Reference	Contaminated	Reference	Contaminated
	3.5	2.6	0.54	0.41
	5.0	18.0	0.70	1.26
	36.0	110.0	1.56	2.04
	68.0	1300.0	1.83	3.11
	170.0	6.9	2.23	0.84
	4.3	1.0	0.63	0.00
	7.4	13.0	0.87	1.11
	7.1	1070.0	0.85	3.03
	1.6	661.0	0.20	2.82
	3.8	8.9	0.58	0.95
	35.0	34.0	1.54	1.53
	1.1	24.0	0.04	1.38
	27.0	22.0	1.43	1.34
	19.0	74.0	1.28	1.87
	64.0	80.0	1.81	1.90
	40.0	1900.0	1.60	3.28
	320.0	2.4	2.51	0.38
	1.7	1.5	0.23	0.18
	7.8	1.6	0.89	0.20
	1.6	140.0	0.20	2.15
	0.1		-1.30	
	0.1		-1.30	
	2.2		0.34	
	210.0		2.32	
	300.0		2.48	
	1.1		0.04	
	4.0		0.60	
	31.0		1.49	
	7.5		0.88	
	0.1		-1.30	
Mean	46.0	273.5	0.859	1.489
SD	86.5	534.7	1.030	1.025

Turning now to the test of the null hypothesis of equivalence, this again depends on the results of two t-tests. The first test is whether the observed mean difference is significantly lower than -0.301, at the 5% level of significance. As $\bar{d}$ exceeds -0.301 this is clearly not true. The second test is whether the observed mean difference is significantly larger than +0.301, at the 5% level of significance. The test statistic is $(\bar{d} - \mu_{dH})/SE(\bar{d}) = 1.108$, with 48 df. The probability of a value this large or larger is 0.14, so the result is not significant. The two one-sided tests are both non-significant and there is therefore no evidence against the hypothesis that the sites are equivalent.

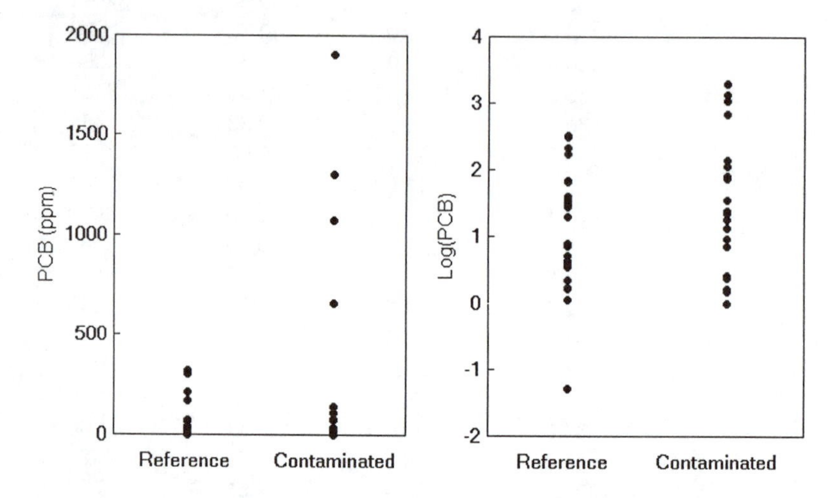

Figure 7.2 The distribution of PCB and $\log_{10}$(PCB) values in a sample of size of size 30 from a reference area and a sample of size 20 from a possibly contaminated area.

The precautionary principle suggests that in a situation like this it is the test of non-equivalence that should be used. It is quite apparent from Gore and Patil's (1994) full set of data that the mean PCB levels are not the same in the phase 1 and the phase 2 sampling areas. Hence the non-significant result for the test of the null hypothesis of equivalence is simply due to the relatively small sample sizes.

Of course, it can reasonably be argued that this example is not very sensible because if the mean PCB concentration is lower in the potentially damaged area then no one would mind. This suggests that one-sided tests are needed rather than the two-sided tests presented here. From this point of view, this example should just be regarded as an illustration of the TOST calculations, rather than what might be done in practice.

7.5 Chapter Summary

- Classical null hypothesis tests may not be appropriate in situations such as deciding whether an impacted site has been reclaimed because the initial assumption should be that this is not the case. The null hypothesis should be that the site is still impacted.

- The United States Environmental Protection Agency recommends that for a site that has not been declared impacted the null hypothesis should be that this is true and the alternative hypothesis should be that an impact has occurred. These hypotheses are reversed for a site that has been declared to be impacted.

- An alternative to a usual hypothesis test involves testing for bioequivalence (two sites are similar enough to be considered equivalent for practical purposes). For example, the test could be of the hypothesis that the density of plants at the impacted site is at least 80% of the density at a control site.

- With two-sided situations, where a reclaimed site should not have a mean that is either too high or too low, the simplest approach for testing for bioequivalence is called the two one-sided test (TOST) that was developed for testing the bioequivalence of two drugs. There are two versions of this that are described. The first version, in line with the precautionary principle (a site is considered to be damaged until there is real evidence to the contrary), has the null hypothesis that the two sites are not equivalent (the true mean difference is not within an acceptable range). The second version has the null hypothesis that the two sites are equivalent.

- Bioequivalence can be defined in terms of the ratio of the means at two sites if this is desirable.

- The two approaches for assessing bioequivalence in terms of an allowable range of mean differences are illustrated using data on PCB concentrations at the Armagh compressor station located in Pennsylvania.

CHAPTER 8

Time Series Analysis

8.1 Introduction

Time series have had a role to play in several of the earlier chapters. In particular, environmental monitoring (Chapter 5) usually involves collecting observations over time at some fixed sites, so that there is a time series for each of these sites, and the same is true for impact assessment (Chapter 6). However, the emphasis in the present chapter will be different, because the situations that will be considered are where there is a single time series, which may be reasonably long (say with 50 or more observations) and the primary concern will often be to understand the structure of the series.

There are several reasons why a time series analysis may be important. For example:

- It gives a guide to the underlying mechanism that produces the series.

- It is sometimes necessary to decide whether a time series displays a significant trend, possibly taking into account serial correlation which, if present, can lead to the appearance of a trend in stretches of a time series although in reality the long-run mean of the series is constant.

- A series shows seasonal variation through the year which needs to be removed in order to display the true underlying trend.

- The appropriate management action depends on the future values of a series, so it is desirable to forecast these and understand the likely size of differences between the forecast and true values.

There is a vast literature on the modelling of time series. It is not possible to cover this in any detail here, so what is done is just to provide an introduction to some of the more popular types of models, and provide references to where more information can be found.

8.2 Components of Time Series

To illustrate the types of time series that arise, some examples can be considered. The first is Jones *et al.*'s (1998a,b) temperature reconstructions for the northern and southern hemispheres, 1000 to 1991 AD. These two series were constructed using data on temperature-sensitive proxy variables including tree rings, ice cores, corals, and historic documents, from 17 sites worldwide. They are plotted in Figure 8.1.

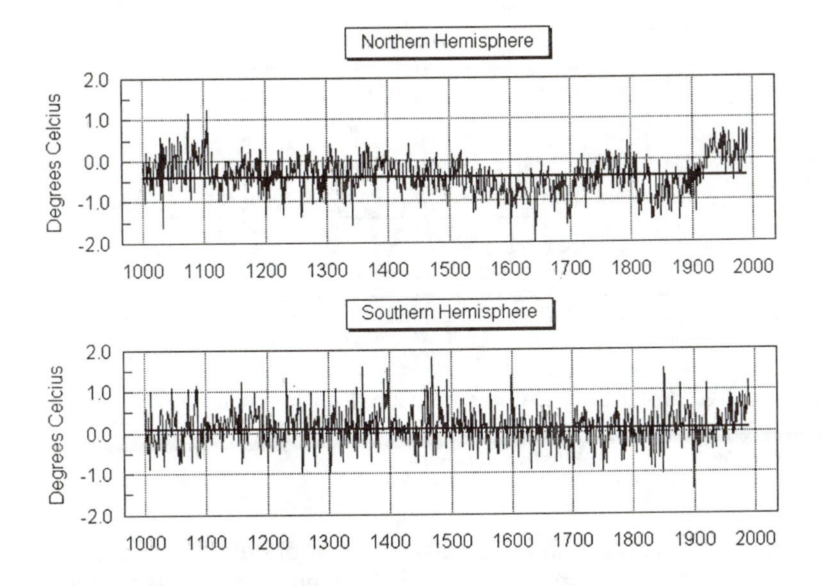

Figure 8.1 Average northern and southern hemisphere temperature series 1000 to 1991 AD calculated by Jones *et al.* (1998a,b) using data from temperature-sensitive proxy variables at 17 sites worldwide. The heavy horizontal lines on each plot are the overall mean temperatures.

The series are characterised by a considerable amount of year to year variation, with excursions away from the overall mean for periods up to about 100 years, with these excursions being more apparent in the northern hemisphere series. The excursions are typical of the behaviour of series with a fairly high level of serial correlation.

In view of the current interest in global warming it is interesting to see that the northern hemisphere temperatures in the latter part of the present century are warmer than the overall mean, but similar to those seen in the latter part of the tenth century, although somewhat less variable. The recent pattern of warm southern hemisphere temperatures is not seen earlier in the series.

A second example is a time series of the water temperature of a stream in Dunedin, New Zealand, measured every month from January 1989 to December 1997. The series is plotted in Figure 8.2. In this case, not surprisingly, there is a very strong seasonal component, with the warmest temperatures in January to March, and the coldest temperatures in about the middle of the year. There is no clear trend, although the highest recorded temperature was in January 1989, and the lowest was in August 1997.

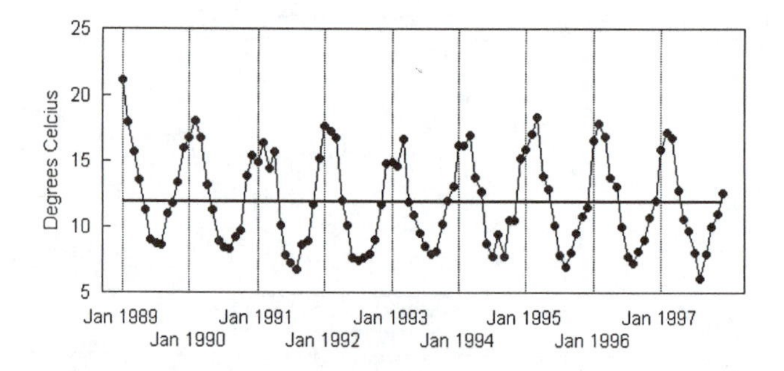

Figure 8.2 Water temperatures measured on a stream in Dunedin, New Zealand, at monthly intervals from January 1989 to December 1997. The overall mean is the heavy horizontal line.

A third example is the estimated number of pairs of the sandwich tern (*Sterna sandvicenis*) on Dutch Wadden Island, Griend, for the years 1964 to 1995, as provided by Schipper and Meelis (1997). The situation is that in the early 1960s the number of breeding pairs decreased dramatically because of poisoning by chlorated hydrocarbons. The discharge of these toxicants was stopped in 1964, and estimates of breeding pairs were then made annually to see whether numbers increased. Figure 8.3 shows the estimates obtained.

The time series in this case is characterised by an upward trend, with substantial year to year variation around this trend. Another point to note is that the year to year variation increased as the series increased. This is an effect that is frequently observed in series with a strong trend.

Finally, Figure 8.4 shows yearly sunspot numbers from 1700 to the present (Sunspot Index Data Center, 1999). The most obvious characteristic of this series is the cycle of about 11 years, although it is also apparent that the maximum sunspot number varies considerably from cycle to cycle.

The examples demonstrate the types of components that may appear in a time series. These are:

(a) a trend component, such that there is a long-term tendency for the values in the series to increase or decrease (as for the sandwich tern);

(b) a seasonal component for series with repeated measurements within calendar years, such that observations at certain times of the year tend to be higher or lower than those at certain other times of the year (as for the water temperatures in Dunedin);

(c) a cyclic component that is not related to the seasons of the year (as for sunspot numbers);

(d) a component of excursions above or below the long-term mean or trend that is not associated with the calendar year (as for global temperatures); and

(e) a random component affecting individual observations (as in all the examples).

These components cannot necessarily be separated easily. For example, it may be a question of definition as to whether the component (d) is part of the trend in a series, or is a deviation from the trend.

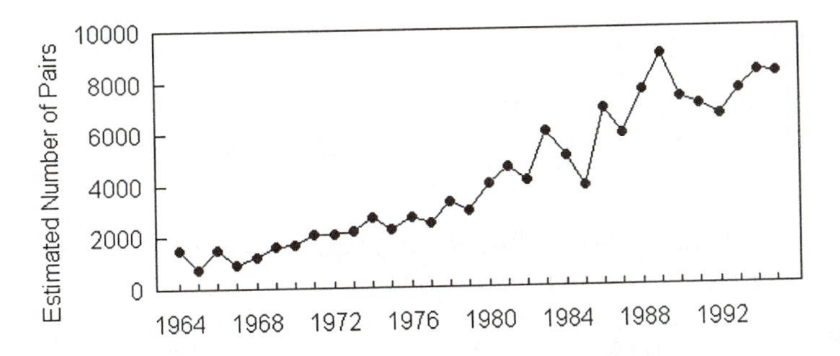

Figure 8.3 The estimated number of breeding sandwich tern pairs on the Dutch Wadden Island, Griend, from 1964 to 1995.

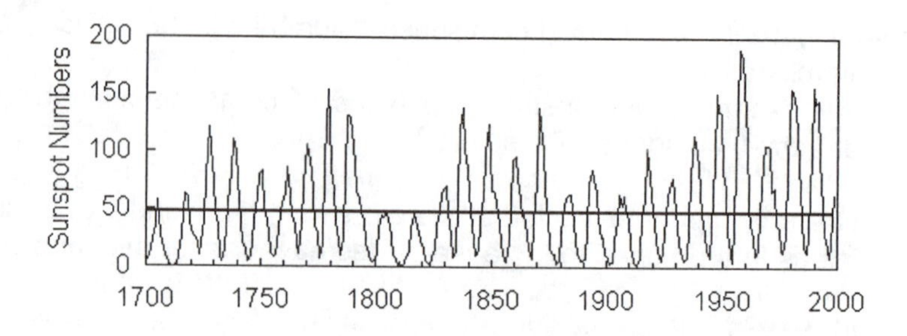

Figure 8.4 Yearly sunspot numbers since 1700 from the Sunspot Index Data Center maintained by the Royal Observatory of Belgium.

8.3 Serial Correlation

Serial correlation coefficients measure the extent to which the observations in a series separated by different time differences tend to be similar. They are calculated in a similar way to the usual Pearson correlation coefficient between two variables. Given data (x_1,y_1), (x_2,y_2), ..., (x_n,y_n) on n pairs of observations for variables X and Y, the sample Pearson correlation is calculated as

$$r = \sum_{i=1}^{n} (x_i - \bar{x})(y_i - \bar{y}) / \sqrt{[\sum_{i=1}^{n} (x_i - \bar{x})^2 \sum_{i=1}^{n} (y_i - \bar{y})^2]}, \quad (8.1)$$

where $\bar{x}$ is the sample mean for X and $\bar{y}$ is the sample mean for Y.

Equation (8.1) can be applied directly to the values (x_1,x_2), (x_2,x_3), ..., (x_{n-1},x_n) in a time series to estimate the serial correlation, r_1, between terms one time period apart. However, what is usually done is to calculate this using a simpler equation, such as

$$r_1 = [\sum_{i=1}^{n-1} (x_i - \bar{x})(x_{i+1} - \bar{x})/(n-1)]/[\sum_{i=1}^{n} (x_i - \bar{x})^2]/n], \quad (8.2)$$

where $\bar{x}$ is the mean of the whole series. Similarly, the correlation between x_i and x_{i+k} can be estimated by

$$r_k = [\sum_{i=1}^{n-k} (x_i - \bar{x})(x_{i+k} - \bar{x}) / (n-k)] / [\sum_{i=1}^{n} (x_i - \bar{x})^2 / n]. \quad (8.3)$$

This is sometimes called the autocorrelation at lag k.

There are some variations on equations (8.2) and (8.3) that are sometimes used, and when using a computer program it may be necessary to determine what is actually calculated. However, for long time series the different varieties of equations give almost the same values.

The correlogram, which is also called the autocorrelation function (ACF), is a plot of the serial correlations r_k against k. It is a useful diagnostic tool for gaining some understanding of the type of series that is being dealt with. A useful result in this respect is that if a series is not too short (say $n > 40$) and consists of independent random values from a single distribution (i.e., there is no autocorrelation), then the statistic r_k will approximately normally distributed with a mean of

$$E(r_k) \approx -1/(n - 1), \tag{8.4}$$

and a variance of

$$\mathrm{Var}(r_k) \approx 1/n. \tag{8.5}$$

The significance of the sample serial correlation r_k can therefore be assessed by seeing whether it falls within the limits $-1/(n - 1) \pm 1.96/\sqrt{n}$. If it is within these limits, then it is not significantly different from 0 at about the 5% level.

Note that there is a multiple testing problem here because if r_1 to r_{20} are all tested at the same time, for example, then one of these values can be expected to be significant by chance (Section 4.9). This suggests that the limits $-1/(n - 1) \pm 1.96/\sqrt{n}$ should be used only as a guide to the importance of serial correlation, with the occasional value outside the limits not being taken too seriously.

Figures 8.5 shows the correlograms for the global temperature time series (Figure 8.1). It is interesting to see that these are quite different for the northern and southern hemisphere temperatures. It appears that for some reason the northern hemisphere temperatures are significantly correlated even up to about 70 years apart in time. However, the southern hemisphere temperatures show little correlation after they are two years or more apart in time.

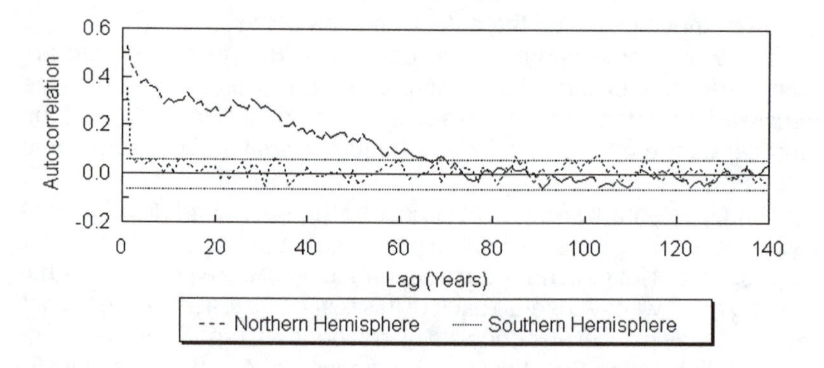

Figure 8.5 Correlograms for northern and southern hemisphere temperatures, 1000 to 1991 AD, with the broken horizontal lines indicating the limits within which autocorrelations are expected to lie 95% of the time for random series of this length.

Figure 8.6 shows the correlogram for the series of monthly temperatures measured for a Dunedin stream (Figure 8.2). Here the effect of seasonal variation is very apparent, with temperatures showing high but decreasing correlations for 12, 24, 36 and 48 month time lags.

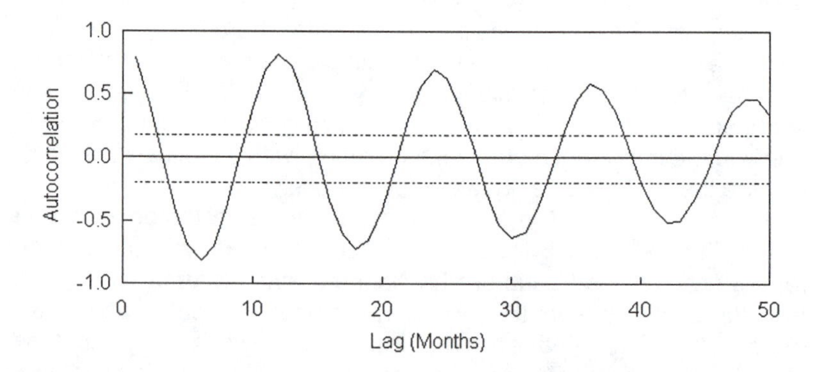

Figure 8.6 Correlogram for the series of monthly temperatures in a Dunedin stream, with the broken horizontal lines indicating the limits on autocorrelations expected for a random series of this length.

The time series of the estimated number of pairs of the sandwich tern on Wadden Island displays increasing variation as the mean increases (Figure 8.3). However, the variation is more constant if the logarithm to base 10 of the estimated number of pairs is considered (Figure 8.7). The correlogram has therefore been calculated for the logarithm series, and this is shown in Figure 8.8. Here the autocorrelation is high for observations one year apart, decreases to about -0.4 for observations 22 years apart, and then starts to increase again. This pattern must be largely due to the trend in the series.

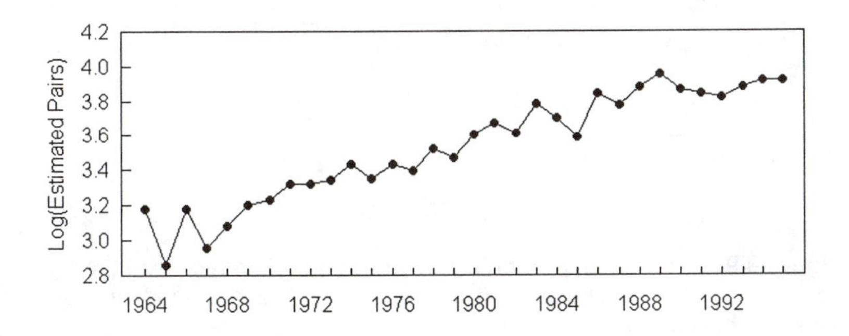

Figure 8.7 Logarithms (base 10) of the estimated number of pairs of the sandwich tern at Wadden Island.

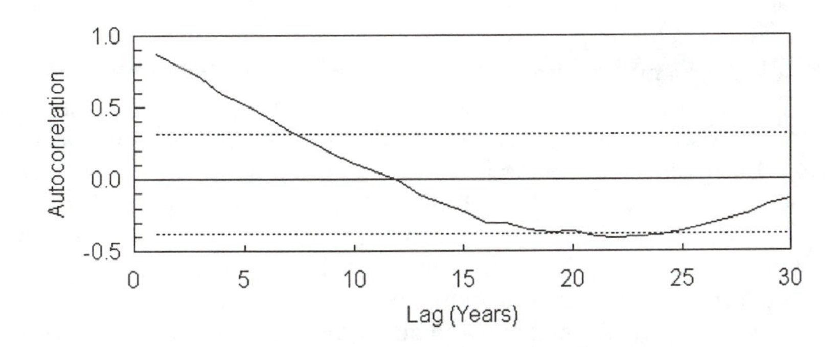

Figure 8.8 Correlogram for the series of logarithms of the number of pairs of sandwich terns on Wadden Island, with the broken horizontal lines indicating the limits on autocorrelations expected for a random series of this length.

Finally, the correlogram for the sunspot numbers series (Figure 8.4) is shown in Figure 8.9. The 11 year cycle shows up very obviously with high but decreasing correlations for 11, 22, 33 and 44 years. The pattern is similar to what is obtained from the Dunedin stream temperature series with a yearly cycle.

If nothing else, these examples demonstrate how different types of time series exhibit different patterns of structure.

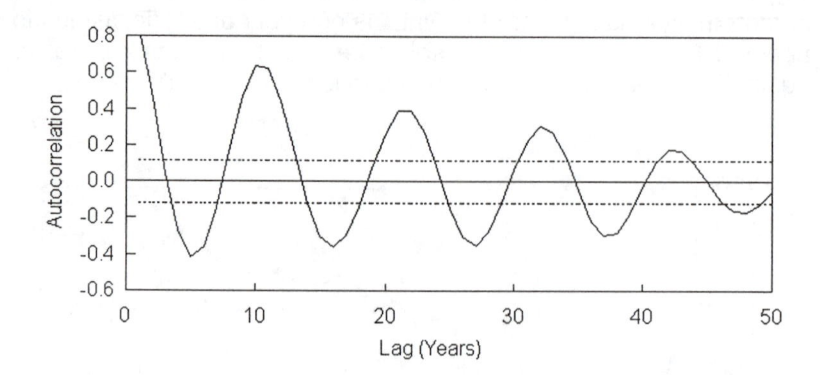

Figure 8.9 Correlogram for the series of sunspot numbers, with the broken horizontal lines indicating the limits on autocorrelations expected for a random series of this length.

8.4 Tests for Randomness

A random time series is one which consists of independent values from the same distribution. There is no serial correlation and this is the simplest type of data that can occur.

There are a number of standard non-parametric tests for randomness that are sometimes included in statistical packages. These may be useful for a preliminary analysis of a time series to decide whether it is necessary to do a more complicated analysis. They are called 'non-parametric' because they are only based on the relative magnitude of observations rather than assuming that these observations come from any particular distribution.

One test is the runs above and below the median test. This involves replacing each value in a series by 1 if it is greater than the median, and 0 if it is less than or equal to the median. The number of runs of the same value is then determined, and compared with the distribution expected if the zeros and ones are in a random order. For example, consider the following series: 1 2 5 4 3 6 7 9 8. The median is 5, so that the series of zeros and ones is 0 0 0 0 0 1 1 1 1. There are M = 2 runs, so this is the test statistic. The trend in the initial series is reflected in M being the smallest possible value. This then needs to be compared with the distribution that is obtained if the zeros and ones are in a random order.

For short series (20 or fewer observations) the observed value of M can be compared with the exact distribution when the null hypothesis is true using tables provided by Swed and Eisenhart (1943), Siegel (1956), or Madansky (1988), among others. For longer series this distribution is approximately normal with mean

$$\mu_M = 2r(n - r)/n + 1, \tag{8.6}$$

and variance

$$\sigma^2_M = 2r(n - r)\{2r(n - r) - n\}/\{n^2(n - 1)\}, \tag{8.7}$$

where r is the number of zeros (Gibbons, 1986, p. 556). Hence

$$Z = (M - \mu_M)/\sigma_M$$

can be tested for significance by comparison with the standard normal distribution (possibly modified with the continuity correction described below).

Another non-parametric test is the sign test. In this case the test statistic is P, the number of positive signs for the differences $x_2 - x_1$, $x_3 - x_2$, ..., $x_n - x_{n-1}$. If there are m differences after zeros have been eliminated, then the distribution of P has mean

$$\mu_P = m/2, \tag{8.8}$$

and variance

$$\sigma^2_P = m/12, \tag{8.9}$$

for a random series (Gibbons, 1986, p. 558). The distribution approaches a normal distribution for moderate length series (say 20 observations or more).

The runs up and down test is also based on the differences between successive terms in the original series. The test statistic is R, the observed number of 'runs' of positive or negative differences. For example, in the case of the series 1 2 5 4 3 6 7 9 8 the signs of the differences are + + - - + + + +, and R = 3. For a random series the mean and variance of the number of runs are

$$\mu_R = (2m+1)/3, \tag{8.10}$$

and

$$\sigma^2_R = (16m-13)/90, \quad (8.11)$$

where m is the number of differences (Gibbons, 1986, p. 557). A table of the distribution is provided by Bradley (1968) among others, and C is approximately normally distributed for longer series (20 or more observations).

When using the normal distribution to determine significance levels for these tests of randomness it is desirable to make a continuity correction to allow for the fact that the test statistics are integers. For example, suppose that there are M runs above and below the median, which is less than the expected number μ_M. Then the probability of a value this far from μ_M is twice the integral of the approximating normal distribution from minus infinity to M + ½, providing that M + ½ is less than μ_M. The reason for taking the integral up to M + ½ rather than M is to take into account the probability of getting exactly M runs, which is approximated by the area from M - ½ to M + ½ under the normal distribution. In a similar way, if M is greater than μ_M then twice the area from M - ½ to infinity is the probability of M being this far from μ_M, providing that M - ½ is greater than μ_M. If μ_M lies within the range from M - ½ to M + ½, then the probability of being this far or further from μ_M is exactly one.

Example 8.1 Minimum Temperatures in Uppsala, 1900 to 1981

To illustrate the tests for randomness just described, consider the data in Table 8.1 for July minimum temperatures in Uppsala, Sweden, for the years 1900 to 1981. This is part of a long series started by Anders Celsius, the Professor of Astronomy at the University of Uppsala, who started collecting daily measurements in the early part of the eighteenth century. There are almost complete daily temperatures from the year 1739, although true daily minimums are only recorded from 1839 when a maximum-minimum thermometer started to be used (Jandhyala *et al.*, 1999). Minimum temperatures in July are recorded by Jandhyala *et al.* for the years 1774 to 1981, as read from a figure given by Leadbetter *et al.* (1983), but for the purpose of this example only the last part of the series is tested for randomness.

A plot of the series is shown in Figure 8.10. The temperatures were low in the early part of the century, but then increased and became fairly constant.

The number of runs above and below the median is M = 42. From equations (8.6) and (8.7) the expected number of runs for a random series is also $\mu_M = 42.0$, with standard deviation $\sigma_M = 4.50$. Clearly, this is not a significant result. For the sign test, the number of positive differences is P = 44, out of m = 81 non-zero differences. From equations (8.8) and (8.9) the mean and standard deviation for P for a random series are $\mu_P = 40.5$ and $\sigma_P = 2.6$. With the continuity correction described above, the significance can be determined by comparing $Z = (P - \frac{1}{2} - \mu_P)/\sigma_P = 1.15$ with the standard normal distribution. The probability of a value this far from zero is 0.25. Hence this gives little evidence of non-randomness. Finally, the observed number of runs up and down is R = 49. From equations (8.10) and (8.11) the mean and standard deviation of R for a random series are $\mu_R = 54.3$ and $\sigma_R = 3.8$. With the continuity correction the observed R corresponds to a score of Z = -1.28 for comparing with the standard normal distribution. The probability of a value this far from zero is 0.20, so this is another insignificant result.

Table 8.1 Minimum July temperatures in Uppsala (°C) for the years 1900 to 1981 (Source: Jandhyala *et al.*, 1999)

Year	Temp	Year	Temp	Year	Temp	Year	Temp	Year	Temp
1900	5.5	1920	8.4	1940	11.0	1960	9.0	1980	9.0
1901	6.7	1921	9.7	1941	7.7	1961	9.9	1981	12.1
1902	4.0	1922	6.9	1942	9.2	1962	9.0		
1903	7.9	1923	6.7	1943	6.6	1963	8.6		
1904	6.3	1924	8.0	1944	7.1	1964	7.0		
1905	9.0	1925	10.0	1945	8.2	1965	6.9		
1906	6.2	1926	11.0	1946	10.4	1966	11.8		
1907	7.2	1927	7.9	1947	10.8	1967	8.2		
1908	2.1	1928	12.9	1948	10.2	1968	7.0		
1909	4.9	1929	5.5	1949	9.8	1969	9.7		
1910	6.6	1930	8.3	1950	7.3	1970	8.2		
1911	6.3	1931	9.9	1951	8.0	1971	7.6		
1912	6.5	1932	10.4	1952	6.4	1972	10.5		
1913	8.7	1933	8.7	1953	9.7	1973	11.3		
1914	10.2	1934	9.3	1954	11.0	1974	7.4		
1915	10.8	1935	6.5	1955	10.7	1975	5.7		
1916	9.7	1936	8.3	1956	9.4	1976	8.6		
1917	7.7	1937	11.0	1957	8.1	1977	8.8		
1918	4.4	1938	11.3	1958	8.2	1978	7.9		
1919	9.0	1939	9.2	1959	7.4	1979	8.1		

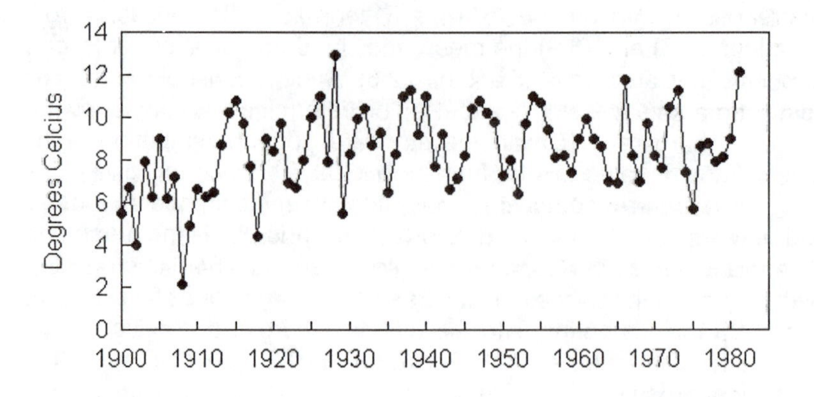

Figure 8.10 Minimum July temperatures in Uppsala, Sweden, for the years 1900 to 1981.

None of the non-parametric tests for randomness give any evidence against this hypothesis, even though it appears that the mean of the series was lower in the early part of the century than it has been more recently. This suggests that it is also worth looking at the correlogram, which indicates some correlation in the series from one year to the next. But even here the evidence for non-randomness is not very marked (Figure 8.11). The question of whether the mean was constant for this series is considered again in the next section.

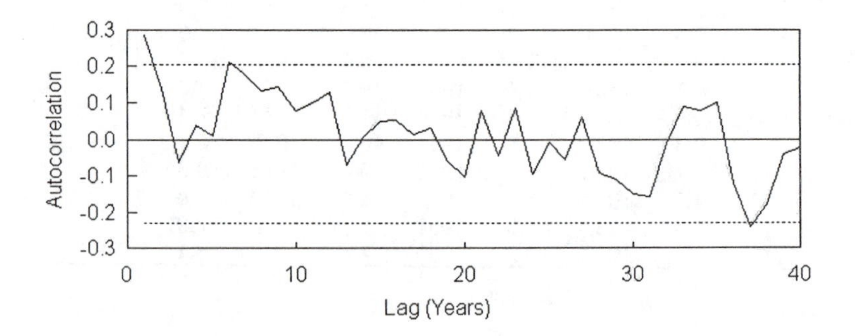

Figure 8.11 Correlogram for the minimum July temperatures in Uppsala, with the 95% limits on autocorrelations for a random series shown by the broken horizontal lines.

8.5 Detection of Change Points and Trends

Suppose that a variable is observed at a number of points of time, to give a time series x_1, x_2, ..., x_n. The change point problem is then to detect a change in the mean of the series if this has occurred at an unknown time between two of the observations. The problem is much easier if the point where a change might have occurred is know, which then requires what is sometimes called an intervention analysis.

A formal test for the existence of a change point seems to have first been proposed by Page (1955) in the context of industrial process control. Since that time a number of other approaches have been developed, as reviewed by Jandhyala and MacNeill (1986), and Jandhyala *et al.* (1999). Methods for detecting a change in the mean of an industrial process through control charts and related techniques have been considerably developed (Sections 5.7 and 5.8). Bayesian methods have also been investigated (Carlin *et al.*, 1992), and Sullivan and Woodhall (1996) suggest a useful approach for examining data for a change in the mean and/or the variance at an unknown time.

The main point to note about the change point problem is that it is not valid to look at the time series, decide where a change point may have occurred, and then test for a significant difference between the means for the observations before and after the change. This is because the maximum mean difference between two parts of the time series may be quite large by chance alone and is liable to be statistically significant if it is tested ignoring the way that it was selected. Some type of allowance for multiple testing (Section 4.9) is therefore needed. See the references given above for details of possible approaches.

A common problem with an environmental time series is the detection of a monotonic trend. Complications include seasonality and serial correlation in the observations. When considering this problem it is most important to define the time scale that is of interest. As pointed out by Loftis *et al.* (1991), in most analyses that have been conducted in the past there has been an implicit assumption that what is of interest is a trend over the time period for which data happen to be available. For example, if 20 yearly results are known, then a 20 year trend has implicitly been of interest. This then means that an increase in the first ten years followed by a decrease in the last ten years to the original level has been considered to give no overall trend, with the intermediate changes possibly being thought of as due to serial correlation. This is clearly not appropriate if systematic changes over a five year period (say) are thought of by managers as being 'trend'.

When serial correlation is negligible, regression analysis provides a very convenient framework for testing for trend. In simple cases, a regression of the measured variable against time will suffice, with a test to see whether the coefficient of time is significantly different from zero. In more complicated cases there may be a need to allow for seasonal effects and the influence of one or more exogenous variables. Thus, for example, if the dependent variable is measured monthly, then the type of model that might be investigated is

$$Y_t = ß_1 M_{1t} + ß_2 M_{2t} + \ldots + ß_{12} M_{12t} + \alpha X_t + \theta t + \epsilon_t, \qquad (8.12)$$

where Y_t is the observation at time t, M_{kt} is a month indicator that is 1 when the observation is for month k or is otherwise 0, X_t is a relevant covariate measured at time t, and ϵ_t is a random error. Then the parameters $ß_1$ to $ß_{12}$ allow for differences in Y values related to months of the year, the parameter α allows for an effect of the covariate, and θ is the change in Y per month after adjusting for any seasonal effects and effects due to differences in X from month to month. There is no separate constant term because this is incorporated by the allowance for month effects. If the estimate of θ obtained by fitting the regression equation is significant, then this provides the evidence for a trend.

A small change can be made to the model in order to test for the existence of seasonal effects. One of the month indicators (say the first or last) can be omitted from the model and a constant term introduced. A comparison between the fit of the model with just a constant and the model with the constant and month indicators then shows whether the mean value appears to vary from month to month.

If a regression equation such as the one above is fitted to data, then a check for serial correlation in the error variable ϵ_{ij} should always be made. The usual method involves using the Durbin-Watson test (Durbin and Watson, 1951), for which the test statistic is

$$V = \sum_{i=2}^{n} (e_i - e_{i-1}) / \sum_{i=1}^{n} e_i^2, \qquad (8.13)$$

where there are n observations altogether, and e_1 to e_n are the regression residuals in time order. The expected value of V is 2 when there is no autocorrelation. Values less than 2 indicate a tendency for observations that are close in time to be similar (positive autocorrelation), and values greater than 2 indicate a tendency for close observations to be different (negative autocorrelation).

Table B5 can be used to assess the significance of an observed value of V for a two-sided test at the 5% level. The test is a little unusual as there are values of V that are definitely not significant, values where the significance is uncertain, and values that are definitely significant. This is explained with the table. The Durbin-Watson test does assume that the regression residuals are normally distributed. It should therefore be used with caution if this does not seem to be the case.

If autocorrelation seems to be present, then the regression model can still be used. However, it should be fitted using a method that is more appropriate than ordinary least-squares. Edwards and Coull (1987), Judge *et al.* (1988, pp. 388-93 and 532-8), Neter *et al.* (1983, Chapter 13) and Zetterqvist (1991) all describe how this can be done. Some statistical packages provide one or more of these methods as options. One simple approach is described in Section 8.6 below.

Actually, some researchers have tended to favour non-parametric tests for trend because of the need to analyse large numbers of series with a minimum amount of time devoted to considering the special needs of each series. Thus transformations to normality, choosing models, etc. are to be avoided if possible. The tests for randomness that have been described in the previous section are possibilities in this respect, with all of them being sensitive to trends to some extent. However, the non-parametric methods that currently appear to be most useful are the Mann-Kendall test, the seasonal Kendall test, and the seasonal Kendall test with a correction for serial correlation (Taylor and Loftis, 1989; Harcum *et al.*, 1992).

The Mann-Kendall test is appropriate for data that do not display seasonal variation, or for seasonally corrected data, with negligible autocorrelation. For a series $x_1, x_2, ..., x_n$ the test statistic is the sum of the signs of the differences between all pairwise observations,

$$S = \sum_{i=2}^{n} \sum_{j=1}^{i-1} \text{sign}(x_i - x_j), \qquad (8.14)$$

where sign(z) is -1 for $z < 0$, 0 for $z = 0$, and +1 for $z > 0$. For a series of values in a random order the expected value of S is zero and the variance is

$$\text{Var}(S) = n(n - 1)(2n + 5)/18. \qquad (8.15)$$

To test whether S is significantly different from zero it is best to use a special table if n is ten or less (Helsel and Hirsch, 1992, p. 469). For larger values of n, $Z_S = S/\sqrt{\text{Var}(S)}$ can be compared with critical values for the standard normal distribution.

To accommodate seasonality in the series being studied, Hirsch *et al.* (1982) introduced the seasonal Kendall test. This involves calculating the statistic S separately for each of the seasons of the year (weeks, months, etc.) and uses the sum for an overall test. Thus if S_j is the value of S for season j, then on the null hypothesis of no trend S_T =

$\sum S_j$ has an expected value of 0 and a variance of $Var(S_T) = \sum Var(S_j)$. The statistic $Z_T = S_T/\sqrt{Var(S_T)}$ can therefore be used for an overall test of trend by comparing it with the standard normal distribution. Apparently the normal approximation is good providing that the total series length is 25 or more.

An assumption with the seasonal Kendall test is that the statistics for the different seasons are independent. When this is not the case an adjustment for serial correlation can be made when calculating $Var(\sum S_T)$ (Hirsch and Slack, 1984; Zetterqvist, 1991). An allowance for missing data can also be made in this case.

Finally, an alternative approach for estimating the trend in a series without assuming that it can be approximated by a particular parametric function involves using a moving average type of approach. Computer packages often offer this type of approach as an option, and more details can be found in specialist texts on time series analysis (e.g., Chatfield, 1989, p. 13).

Example 8.2 Minimum Temperatures in Uppsala, Reconsidered

In Example 8.1 tests for randomness were applied to the data in Table 8.1 on minimum July temperatures in Uppsala for the years 1900 to 1981. None of the tests gave evidence for non-randomness, although some suggestion of autocorrelation was found. The non-significant results seem strange because the plot of the series (Figure 8.10) gives an indication that the minimum temperatures tended to be low in the early part of the century. In this example the series is therefore reexamined, with evidence for changes in the mean level of the series being specifically considered.

First, consider a regression model for the series, of the form

$$Y_t = \beta_0 + \beta_1 t + \beta_2 t^2 + \ ... + \beta_p t^p + \epsilon_t, \qquad (8.16)$$

where Y_t is the minimum July temperature in year t, taking t = 1 for 1900 and t = 82 for 1981, and ϵ_t is a random deviation from the value given by the polynomial for the year t. Trying linear, quadratic, cubic and quartic models gives the analysis of variance shown in Table 8.2. It is found that the linear and quadratic terms are highly significant, the cubic term is not significant at the 5% level, and the quartic term is not significant at all. A quadratic model therefore seems appropriate.

Table 8.2 Analysis of variance for polynomial trend models fitted to the time series of minimum July temperatures in Uppsala

Source of variation	Sum of squares	Degrees of freedom	Mean square	F-ratio	Significance (p-value)
Time	31.64	1	31.64	10.14	0.002
$Time^2$	29.04	1	29.04	9.31	0.003
$Time^3$	9.49	1	9.49	3.04	0.085
$Time^4$	2.23	1	2.23	0.71	0.400
Error	240.11	77	3.12		
Total	312.51	81			

When a simple regression model like this is fitted to a time series it is most important to check that the estimated residuals do not display autocorrelation. The Durbin-Watson statistic of equation (8.13) is $V = 1.69$ for this example. With $n = 82$ observations and $p = 2$ regression variables, Table B5 shows that to be definitely significant at the 5% level on a two-sided test V would have to be less than about 1.52. Hence there is little concern about autocorrelation for this example.

Figure 8.12 shows plots of the original series, the fitted quadratic trend curve, and the standardized residuals (the differences between the observed temperature and the fitted trend values divided by the estimated residual standard deviation). The model appears to be a very satisfactory description of the data, with the expected temperature appearing to increase from 1900 to 1930 and then remain constant, or even decline slightly. The residuals from the regression model are approximately normally distributed, with almost all of the standardized residuals in the range from -2 to +2.

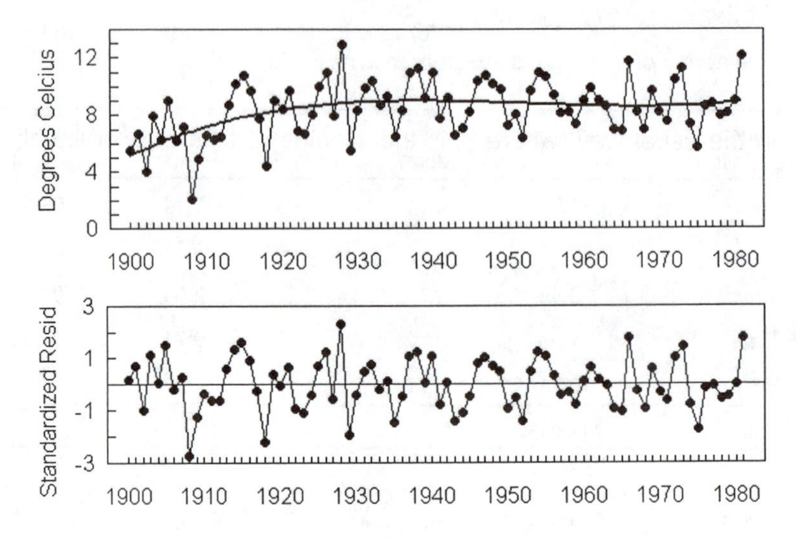

Figure 8.12 A quadratic trend fitted to the series of minimum July temperatures in Uppsala (top graph), with the standardized residuals from the fitted trend (lower graph).

The Mann-Kendall test using the statistic S calculated using equation (8.14) also gives strong evidence of a trend in the mean of this series. The observed value of S is 624, with a standard deviation of 249.7. The Z-score for testing significance is therefore Z = 624/249.7 = 2.50, and the probability of a value this far from zero is 0.012 for a random series.

See Smith (1993) for a discussion of other approaches for the analysis of long-term weather data.

8.6 More Complicated Time Series Models

The internal structure of time series can mean that quite complicated models are needed to describe the data. No attempt will be made here to cover the huge literature on this topic. Instead, the most commonly used types of models will be described, with some simple examples of their use. For more information a specialist text such as that of Chatfield (1989) should be consulted.

The possibility of allowing for autocorrelation with a regression model was mentioned in the last section. Assuming the usual regression situation where there are n values of a variable Y and corresponding values for variables X_1 to X_p, one simple approach that can be used is as follows:

(a) Fit the regression model

$$y_t = \beta_0 + \beta_1 x_{1t} + \ldots + \beta_p x_{pt} + \epsilon_t,$$

in the usual way, where y_t is the Y value at time t, for which the values of X_1 to X_p are x_{1t} to x_{pt}, respectively, and ϵ_t is the error term. Let the estimated equation be

$$\hat{y} = b_0 + b_1 x_{1t} + \ldots + b_p x_{pt},$$

with estimated regression residuals

$$e_t = (y_t - \hat{y}_t).$$

(b) Assume that the residuals in the original series are correlated because they are related by an equation of the form

$$\epsilon_t = \alpha\epsilon_{t-1} + u_t,$$

where α is a constant and u_t is a random value from a normal distribution with mean zero and a constant standard deviation. Then it turns out that α can be estimated by $\hat{\alpha}$, the first order serial correlation for the estimated regression residuals.

(c) Note that from the original regression model

$$y_t - \alpha y_{t-1} = \beta_0(1 - \alpha) + \beta_1(x_{1t} - \alpha x_{1t-1}) + \ldots + \beta_p(x_{pt} - \alpha x_{pt-1}) + \epsilon_t - \alpha\epsilon_{t-1}$$

or

$$z_t = \gamma + \beta_1 v_{1t} + \ldots + \beta_p v_{pt} + u_t,$$

where $z_t = y_t - \alpha y_{t-1}$, $\gamma = \beta_0(1 - \alpha)$, and $v_{it} = x_{it} - \alpha x_{it-1}$, for i = 1, 2, ..., p. This is now a regression model with independent errors, so that the coefficients γ and β_1 to β_p can be estimated by ordinary regression, with all the usual tests of significance, etc. Of course, α is not known. The approximate procedure actually used is therefore to replace α with the estimate $\hat{\alpha}$ for the calculation of the z_t and v_{it} values.

These calculations can be done easily enough in most statistical packages. A variation called the Cochran-Orcutt procedure involves iterating using steps (b) and (c). What is then done is to recalculate the regression residuals using the estimates of β_0 to β_p obtained at step (c), obtain a new estimate of α using these, and then repeat step (c). This is continued until the estimate of α becomes constant to a few decimal places. Another variation that is available in some computer packages is to estimate α and β_0 to β_1 simultaneously using maximum likelihood.

The validity of this type of approach depends on the simple model $\epsilon_t = \alpha\epsilon_{t-1} + u_t$ being reasonable to account for the autocorrelation in the regression errors. This can be assessed by looking at the correlogram for the series of residuals $e_1, e_2, ..., e_n$ calculated at step (a) of the above procedure. If the autocorrelations appear to decrease quickly to approximately zero with increasing lags, then the assumption is probably reasonable. This is the case, for example, with the correlogram for southern hemisphere temperatures, but less obviously true for northern hemisphere temperatures.

The model $\epsilon_t = \alpha\epsilon_{t-1} + u_t$ is the simplest type of autoregressive model. In general these models take the form

$$x_t = \mu + \alpha_1(x_{t-1} - \mu) + \alpha_2(x_{t-2} - \mu) + ... + \alpha_p(x_{t-p} - \mu) + \epsilon_t, \qquad (8.17)$$

where μ is the mean of the series, α_1 to α_p are constants, and ϵ_t is an error term with a constant variance which is independent of the other terms in the model. This type of model, which is sometimes called AR(p), may be reasonable for series where the value at time t depends only on the previous values of the series plus random disturbances that are accounted for by the error terms. Restrictions are required on the α values to ensure that the series is stationary, which means in practice that the mean, variance, and autocorrelations in the series are constant with time.

To determine the number of terms that are required in an autoregressive model, the partial autocorrelation function (PACF) is useful. The pth partial autocorrelation shows how much of the correlation in a series is accounted for by the term $\alpha_p(x_{t-p} - \mu)$ in the model of equation (8.17), and its estimate is just the estimate of the coefficient α_p.

Moving average models are also commonly used. A time series is said to be a moving average process of order q, MA(q), if it can be described by an equation of the form

$$x_t = \mu + \beta_0 z_t + \beta_1 z_{t-1} + ... + \beta_q z_{t-q}, \qquad (8.18)$$

where the values of $z_1, z_2, ..., z_t$ are random values from a distribution with mean zero and constant variance. Such models are useful when the autocorrelation in a series drops to close to zero for lags of more than q.

Mixed autoregressive-moving average (ARMA) models combine the features of equations (8.17) and (8.18). Thus a ARMA(p,q) model takes the form

$$x_t = \mu + \alpha_1(x_{t-1} - \mu) + \ldots + \alpha_p(x_{t-p} - \mu) + \beta_1 z_{t-1} + \ldots + \beta_q z_{t-q}, \quad (8.19)$$

with the terms defined as before. A further generalization is to integrated autoregressive-moving average models (ARIMA), where differences of the original series are taken before the ARMA model is assumed. The usual reason for this is to remove a trend in the series. Taking differences once removes a linear trend, taking differences twice removes a quadratic trend, and so on. Special methods for accounting for seasonal effects are also available with these models.

Fitting these relatively complicated time series to data is not difficult as many statistical packages include an ARIMA option, which can be used either with this very general model, or with the component parts such as autoregression. Using and interpreting the results correctly is another matter, and with important time series it is probably best to get the advice of an expert.

Example 8.3 Temperatures of a Dunedin Stream, 1989 to 1997

For an example of allowing for autocorrelation with a regression model, consider again the monthly temperature readings for a stream in Dunedin, New Zealand. These are plotted in Figure 8.2, and also provided in Table 8.3.

The model that will be assumed for these data is similar to that given by equation (8.12), except that there is a polynomial trend, and no exogenous variable. Thus

$$Y_t = \beta_1 M_{1t} + \beta_2 M_{2t} + \ldots + \beta_{12} M_{12\,t} + \theta_1 t + \theta_2 t^2 + \ldots + \theta_p t^p + \epsilon_t, \quad (8.20)$$

where Y_t is the observation at time t measured in months from 1 in January 1989 to 108 for December 1997, M_{kt} is a month indicator that is 1 when the observation is in month k, where k goes from 1 for January to 12 for December, and ϵ_t is a random error term.

Table 8.3 Monthly temperatures (°C) for a stream in Dunedin, New Zealand for 1989 to 1997

Month	1989	1990	1991	1992	1993	1994	1995	1996	1997
Jan	21.1	16.7	14.9	17.6	14.9	16.2	15.9	16.5	15.9
Feb	17.9	18.0	16.3	17.2	14.6	16.2	17.0	17.8	17.1
Mar	15.7	16.7	14.4	16.7	16.6	16.9	18.3	16.8	16.7
Apr	13.5	13.1	15.7	12.0	11.9	13.7	13.8	13.7	12.7
May	11.3	11.3	10.1	10.1	10.9	12.6	12.8	13.0	10.6
Jun	9.0	8.9	7.9	7.7	9.5	8.7	10.1	10.0	9.7
Jul	8.7	8.4	7.3	7.5	8.5	7.8	7.9	7.8	8.1
Aug	8.6	8.3	6.8	7.7	8.0	9.4	7.0	7.3	6.1
Sep	11.0	9.2	8.6	8.0	8.2	7.8	8.1	8.2	8.0
Oct	11.8	9.7	8.9	9.0	10.2	10.5	9.5	9.0	10.0
Nov	13.3	13.8	11.7	11.7	12.0	10.5	10.8	10.7	11.0
Dec	16.0	15.4	15.2	14.8	13.0	15.2	11.5	12.0	12.5

This model was fitted to the data assuming the existence of first order autocorrelation in the error terms, so that

$$\epsilon_t = \alpha\epsilon_{t-1} + u_t, \tag{8.21}$$

using the maximum likelihood option for autoregression in SPSS (SPSS Inc., 1997). Values of p up to 4 were tried, but there was no significant improvement of the quartic model (p = 4) over the cubic model. Table 8.4 therefore gives the results for the cubic model only. From this table it will be seen that the estimates of θ_1, θ_2 and θ_3 are all highly significantly different from zero. However, the estimate of the autoregressive parameter is not quite significantly different from zero at the 5% level, suggesting that it may not have been necessary to allow for serial correlation at all. On the other hand, it is safer to allow for serial correlation than to ignore it.

The top graph in Figure 8.13 shows the original data, the expected temperatures according to the fitted model, and the estimated trend. Here the estimated trend is the cubic part of the fitted model, which is $-0.140919t + 0.002665t^2 - 0.000015t^3$, with a constant added to make the mean trend value equal to the mean of the original temperature observations. The trend is quite weak, although it is highly significant. The bottom graph shows the estimated u_t values from equation (8.21). These should, and do, appear to be random.

Table 8.4 Estimated parameters for the model of equation (8.20) fitted to the data on monthly temperatures of a Dunedin stream

Parameter	Estimate	Standard error	Ratio[a]	P-value[b]
α (Autoregressive)	0.1910	0.1011	1.89	0.062
β_1 (January)	18.4477	0.5938	31.07	0.000
β_2 (February)	18.7556	0.5980	31.37	0.000
β_3 (March)	18.4206	0.6030	30.55	0.000
β_4 (April)	15.2611	0.6081	25.10	0.000
β_5 (May)	13.3561	0.6131	21.78	0.000
β_6 (June)	11.0282	0.6180	17.84	0.000
β_7 (July)	10.0000	0.6229	16.05	0.000
β_8 (August)	9.7157	0.6276	15.48	0.000
β_9 (September)	10.6204	0.6322	16.79	0.000
β_{10} (October)	11.9262	0.6367	18.73	0.000
β_{11} (November)	13.8394	0.6401	21.62	0.000
β_{12} (December)	16.1469	0.6382	25.30	0.000
θ_1 (Time)	-0.140919	0.04117619	-3.42	0.001
θ_2 (Time^2)	0.002665	0.00087658	3.04	0.003
θ_3 (Time^3)	-0.000015	0.00000529	-2.84	0.006

[a]The estimate divided by the standard error.
[b]Significance level for the ratio when compared with the standard normal distribution. The significance levels do not have much meaning for the month parameters, which all have to be greater than zero.

There is one further check of the model that is worthwhile. This involves examining the correlogram for the u_t series, which should show no significant autocorrelation. The correlogram is shown in Figure 8.14. This is notable for the negative serial correlation of about -0.4 for values six months apart, which is well outside the limits that should contain 95% of values. There is only one other serial correlation outside these limits, for a lag of 44 months, which is presumably just due to chance. The significant negative serial correlation for a lag of six months indicates that the fitted model is still missing some important aspects of the time series. However, overall the model captures most of the structure, and this curious autocorrelation will not be considered further here.

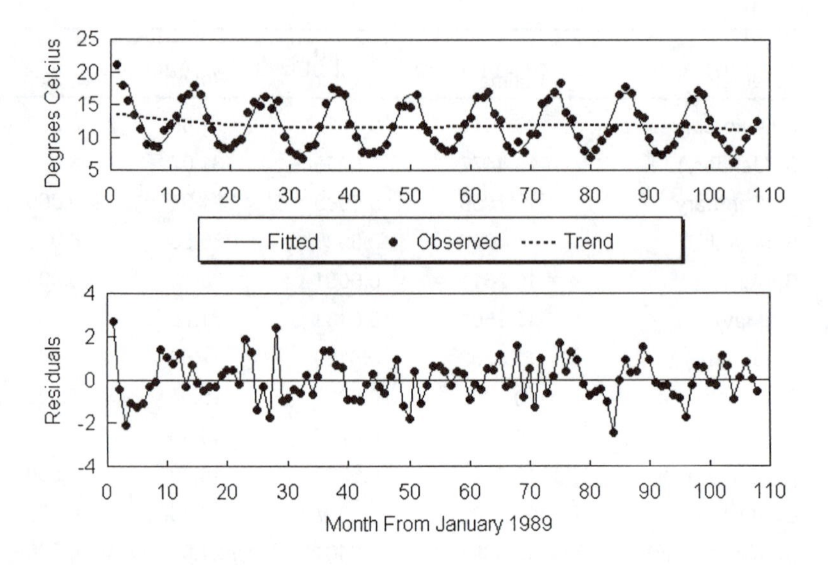

Figure 8.13 The fitted model for the monthly temperature of a Dunedin stream with the estimated trend indicated (top graph), and estimates of the random components u_t in the model (bottom graph).

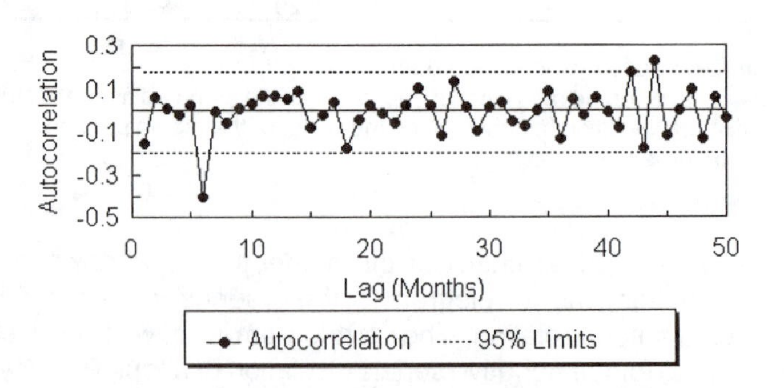

Figure 8.14 The correlogram for the estimated random components u_t in the model for Dunedin stream temperatures. Autocorrelations outside the 95% limits are significantly different from zero at the 5% level.

Example 8.4 Rainfall in Northeast Brazil, 1849 to 1987

For another example, consider the data shown in Table 8.5 and displayed in Figure 8.15. These are yearly rainfall amounts recorded in rain gauges at Fortaleza in the northeast of Brazil, from 1849 to 1987 (Mitchell, 1997). The question to be considered is what is an appropriate model for this series.

The correlogram for the series is shown in Figure 8.16. A first order autoregressive model and other models including more autoregressive terms and moving average terms were fitted to the data using the ARIMA option in SPSS (SPSS Inc., 1997), and it was apparent that the first order autoregressive model is all that is needed. This model was estimated to be

$$x_t = 143.28 + 0.2330(x_{t-1} - 143.28) + \epsilon_t, \tag{8.22}$$

where the autoregressive coefficient of 0.2330 has an estimated standard error of 0.0832, and is highly significant ($t = 2.80$, $p = 0.005$).

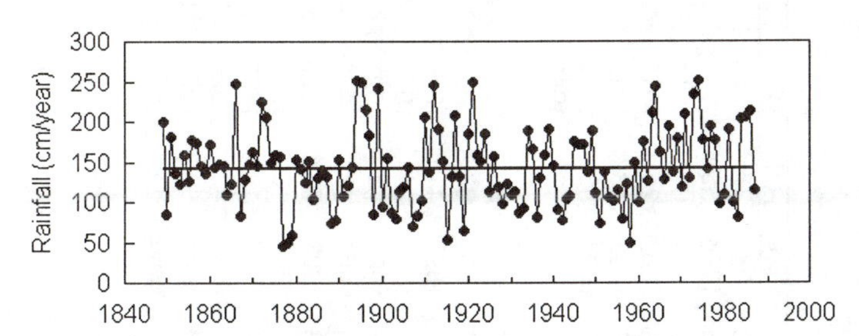

Figure 8.15 Rainfall (cm/year) measured by rain gauges at Fortaleza in northeast Brazil, for the years 1849 to 1987.

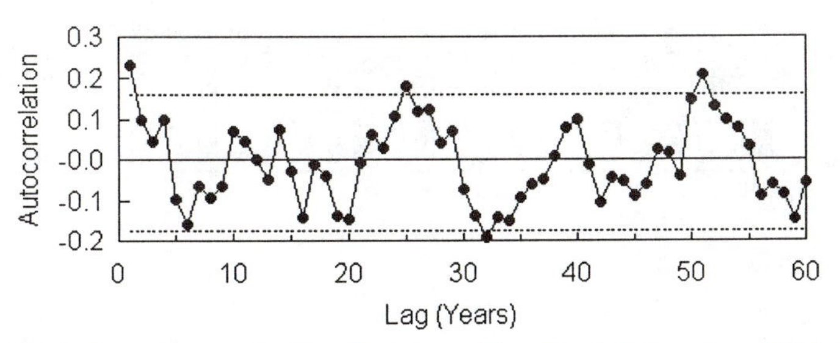

Figure 8.16 Correlogram for the Fortaleza, Brazil, rainfall series, 1849-1987.

Table 8.5 Rainfall (cm/year) measured by rain gauges at Fortaleza in northeast Brazil, for the years 1849 to 1987.

Year	Rain	Year	Rain	Year	Rain	Year	Rain	Year	Rain	Year	Rain	Year	Rain
1849	200.1	1869	147.0	1889	78.4	1909	101.5	1929	123.0	1949	188.1	1969	180.3
1850	85.2	1870	162.8	1890	153.4	1910	205.1	1930	110.7	1950	111.4	1970	119.2
1851	180.6	1871	145.9	1891	107.7	1911	137.3	1931	113.3	1951	74.7	1971	209.3
1852	135.6	1872	225.6	1892	121.1	1912	244.6	1932	87.9	1952	137.8	1972	129.9
1853	123.3	1873	205.8	1893	143.0	1913	190.5	1933	93.7	1953	106.8	1973	233.1
1854	159.0	1874	148.7	1894	250.5	1914	151.2	1934	188.8	1954	103.2	1974	251.2
1855	127.3	1875	158.1	1895	249.1	1915	53.0	1935	166.1	1955	115.2	1975	177.8
1856	177.0	1876	156.9	1896	214.4	1916	132.8	1936	82.0	1956	80.6	1976	141.7
1857	173.4	1877	46.8	1897	183.9	1917	207.7	1937	131.3	1957	122.5	1977	194.1
1858	145.7	1878	50.3	1898	86.3	1918	131.9	1938	158.6	1958	50.4	1978	178.5
1859	135.7	1879	59.7	1899	241.4	1919	65.6	1939	191.1	1959	149.3	1979	98.5
1860	171.6	1880	153.9	1900	94.0	1920	184.7	1940	144.7	1960	101.1	1980	109.5
1861	144.5	1881	142.3	1901	154.5	1921	249.6	1941	91.6	1961	175.9	1981	190.3
1862	146.8	1882	124.6	1902	87.8	1922	159.5	1942	78.0	1962	127.7	1982	99.9
1863	145.2	1883	150.8	1903	78.9	1923	151.3	1943	104.2	1963	211.0	1983	81.6
1864	109.8	1884	104.7	1904	113.6	1924	184.7	1944	109.0	1964	242.6	1984	203.1
1865	123.8	1885	130.7	1905	118.9	1925	113.7	1945	175.0	1965	162.9	1985	206.6
1866	247.8	1886	139.9	1906	143.0	1926	157.1	1946	172.4	1966	128.9	1986	214.0
1867	83.2	1887	132.0	1907	69.7	1927	119.5	1947	172.6	1967	193.7	1987	115.1
1868	128.9	1888	73.6	1908	83.4	1928	99.5	1948	138.4	1968	138.5		

The estimated values for the errors ϵ_t in the model are approximately normally distributed, with no significant serial correlation. The model of equation (8.22) therefore seems quite satisfactory for these data.

8.7 Frequency Domain Analysis

The analyses considered so far in this chapter are called time domain analyses. They are concerned with modelling the observations directly, with models that are intended to explain the components of an observation x_t taken at time t. There is, however, another approach that is used extensively, particularly in areas such as electrical engineering, meteorology, and geophysics. This is called frequency domain analysis, or spectral analysis. Basically, this alternative approach involves attempting to determine how much of the variation in a series is associated with cycles of different lengths. For example, with a 100 year series it is possible to partition the observed variance into components associated with cycles with lengths of 100 years, 50 years, 33.3 years, 25 years, and so on down to 2 years, the cycle lengths being 100/k, for k equal to 1,2, ..., 50.

This type of analysis has obvious benefits if it is cycles in the original series that are of interest. For example, a large part of the variation in the sunspot series (Figure 8.4) is associated with an 11 year cycle. However, many time series analysts find this approach more generally useful for understanding the structure of series.

An introduction to frequency domain analysis of time series is provided by Chatfield (1989, Chapters 6 and 7). Unfortunately, standard statistical packages often fail to provide any options for doing the required calculations so specialist software may need to be obtained by those interested in attempting this approach.

8.8 Forecasting

There are many reasons for carrying out time series analysis. These include an interest in understanding the structure of the series, deciding whether there is evidence of a significant trend, or deciding whether the mean of the series has changed at some point in time due to some outside influence. Often, however, the analysis is conducted in order to be able to make sensible forecasts of future values of the series, preferably with some idea of the size of the forecasting errors that are likely to occur.

Forecasting of a time series using statistical methods involves extrapolating from the known past and present values into the future. Extrapolation is always dangerous, so anyone who attempts this can expect to be seriously wrong at times. With time series the problem is that the model may change for some unexpected reason in the future. Or, indeed, the model may be wrong anyway although it fits the existing data quite well. Actually, models will seldom be exactly correct anyway. Hence extrapolating an approximate model into the future will probably lead to larger forecasting errors than the model itself suggests. Having said all this, the fact is that it is sometimes very useful to have some idea of what the future values of a time series are likely to be. There is therefore a large literature on this topic.

Some of the methods that are often used, and are available in standard statistical packages include the following:

(a) A trend curve is fitted to the available data, and then extrapolated into the future. The trend curve might in this case be linear, a polynomial, or perhaps some more complicated function.

(b) Exponential smoothing is used, where the estimate of the mean of a series at time t is the weighted average of past values given by

$$\hat{x}_t = \alpha x_t + \alpha(1 - \alpha)x_{t-1} + \alpha(1 - \alpha)^2 x_{t-2} + \dots,$$

or equivalently,

$$\hat{x}_t = \alpha x_t + (1 - \alpha)\hat{x}_{t-1}, \qquad (8.23)$$

where α is a smoothing constant with a value between zero and one. The value of α can be chosen by minimizing the prediction errors for the existing data, and equation (8.23) can then be used to predict into the future, with the actual value of x_t replaced by its estimate from the equation. Exponential smoothing can be extended to situations with trend and seasonal variation using something called the Holt-Winters procedure (Chatfield, 1989, p. 70).

(c) An ARIMA model (Section 8.6) can be fitted to the data and then extrapolated forward. This requires the identification of an appropriate model, which is then predicted forward in time. There are some statistical packages that attempt to do this completely automatically.

It would be nice to be able to say which of these methods is generally best. However, this seems impossible because it depends upon the particular circumstances. A detailed discussion of the matters that need to be considered is given by Chatfield (1989, Chapter 5).

8.9 Chapter Summary

- Time series analysis may be important because it gives a guide to the underlying process producing a series. It may be necessary to know whether apparent trends and changes in the mean of a series are real, to remove seasonal variation in order to estimate the underlying changes in a series, or to forecast the future values of a series.

- The components that often occur in time series are a trend (a tendency to increase or decrease), seasonal variation within the calendar year, other cyclic variation, excursions away from the overall mean for relatively short periods of time, and random variation. These components are displayed in a number of examples that are presented.

- Serial correlation coefficients measure the tendency for observations to be similar or different when they are different distances apart in time. If this tendency exists, then it is also called autocorrelation. A plot of the serial correlation coefficients for different distances apart (lags) is called a correlogram.

- A random time series is one in which each of the observations is an independent random value from the same distribution. There are a number of standard non-parametric tests for randomness. The runs above and below the median, the sign test, and the runs up and down test are described, and illustrated on a time series of minimum July temperatures in Uppsala from 1900 to 1981. None of the tests give evidence of non-randomness.

- The change point problem involves determining whether there is any evidence of a change in the mean of a time series, without knowing when the change may have occurred. This requires a proper allowance for multiple testing if all possible times of change are considered.

- The detection of trend in an environmental time series is a common problem. In testing for trend it is important to decide what time scale is important.

- When serial correlation is negligible, regression analysis is a useful tool for detecting trend. Then a regression model is set up which includes an allowance for trend, and the significance of the estimated trend is tested using usual regression methods. The Durbin-Watson test can be used to see whether there is serial correlation in the regression residuals. If there is evidence of serial correlation, then the analysis can be modified to allow for this.

- The Mann-Kendall test is a test for monotonic trend in a series when there is no serial correlation. It can be modified to accommodate seasonality in the series being tested, with a correction for serial correlation between seasons if necessary.

- Some more complicated approaches to time series modelling are described: regression models with serially correlated residuals, autoregressive (AR) models, moving average (MA) models, mixed autoregressive-moving average (ARMA) models, and integrated autoregressive-moving average (ARIMA) models.

- Frequency domain analysis (spectral analysis) is briefly described.

- Methods for forecasting the future values of a time series (extrapolating a fitted trend, exponential smoothing, and the use of ARIMA models) are briefly described.

CHAPTER 9

Spatial Data Analysis

9.1 Introduction

Like time series analysis, spatial data analysis is a highly specialized area in statistics. Therefore, all that is possible in this chapter is to illustrate the types of data that are involved, describe some of the simpler methods of analysis, and give references for where more information can be obtained.

The methods of analysis that are considered in this chapter can be used to:

(a) detect patterns in the locations of objects in space,

(b) quantify correlations between the spatial locations for two types of objects,

(c) measure the spatial autocorrelation for the values of a variable measured over space, and

(d) study the correlation between two variables measured over space when one or both of those variables displays autocorrelation.

To begin with some example sets of data are presented to clarify what exactly (a) to (d) involve.

9.2 Types of Spatial Data

One important category of spatial data is quadrat counts. With such data, the area of interest is divided into many square, rectangular or circular study plots, and the number of objects of interest is counted, in either all of the study plots or a sample of them. Table 9.1 gives an example. In this case the study area is part of a beach near Auckland, New Zealand, the quadrats are circular plots in the sand with an area of 0.1m^2, and the counts are the numbers of two species of shellfish (pipis, *Paphies australis*, and cockles, *Austrovenus stutchburyi*) found down to a depth of 0.15m in the quadrats. These data are a small part of a larger set obtained from a survey to estimate the size of the populations, with some simplification for this example.

Table 9.1 Counts of two species of shellfish from quadrats in an area 200m long by 70m wide of a beach in Auckland, New Zealand. The position of the counts in the table matches the position in the study area for both species, so that the corresponding counts for the two species are from the same quadrat

Distance from	Distance along beach(m)										
low water (m)	0	20	40	60	80	100	120	140	160	180	200
	Counts of pipis (*Paphies australis*)										
0	1	0	4	0	0	0	3	0	2	0	0
10	0	0	0	0	104	0	0	0	1	0	0
20	7	24	0	0	240	0	0	103	1	0	0
30	20	0	0	0	0	0	3	250	7	0	0
40	20	0	2	4	0	222	0	174	4	0	58
50	0	0	11	0	0	126	0	62	7	6	29
60	0	0	7	0	0	0	0	0	23	7	29
70	0	0	0	0	89	0	0	7	8	0	30
	Counts of cockles (*Austrovenus stutchburyi*)										
0	0	0	1	0	0	0	0	0	0	0	0
10	0	0	0	0	7	0	0	0	0	0	0
20	0	0	0	0	9	0	0	3	6	0	0
30	0	0	0	0	0	0	0	0	0	0	0
40	1	0	0	5	0	0	0	7	0	0	10
50	0	0	0	0	0	7	0	10	1	1	19
60	0	0	0	0	0	0	0	0	0	0	2
70	0	0	0	0	2	0	0	2	0	0	16

There are a number of obvious questions that might be asked with these data:

- Are the counts for pipis randomly distributed over the study area, or is there evidence of either a uniform spread or clustering of the higher counts?

- Similarly, are the counts randomly distributed for cockles, or is there evidence of some uniformity or clustering?

- Is there any evidence of an association between the quadrat counts for the two species of shellfish, with the high counts of pipis tending to be in the same quadrat as either high counts or low counts for the cockles?

A similar type of example, but with the data available in a different format is presented in Figure 9.1. Here what is presented are the locations of 45 nests of the ant *Messor wasmani* and 15 nests of the ant *Cataglyphis bicolor* in a 240 by 250 foot area (Harkness and Isham, 1983; Särkkä , 1993, Figure 5.8). *Messor* (species 1) collects seeds for food while *Cataglyphis* (species 2) eats dead insects, mostly *Messor* ants.

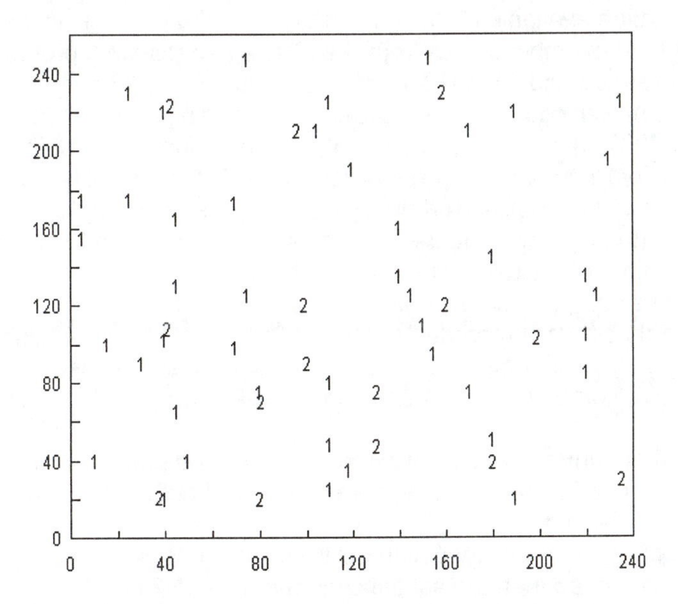

Figure 9.1 Location of 45 nests of *Messor wasmanni* (species 1) and 15 nests of *Cataglyphis bicolor* (species 2) in a 240 ft by 250 ft study area.

Possible questions here are basically the same as for the shellfish:

- Are the positions of the *Messor wasmani* nests randomly located over the study area, or is there evidence of uniformity or clustering in the distribution?

- Similarly, are the *Cataglyphis bicolor* nests random, uniform or clustered in their distribution?

- Is there any evidence of a relationship between the positions of the nests for the two species, for example, with *Messor wasmani* nests tending to be either close to or distant from the *Cataglyphis bicolor* nests?

In this second example there is a point process for each of the two species of ant. It is apparent that this could be made similar to the first example because it would be possible to divide the ant study area into quadrats and compare the two species by quadrat counts, in the same way as for the shellfish. However, knowing the positions of the points instead of just how many are in each quadrat means that there is more information available in the second example than there is in the first.

A third example comes from the Norwegian research programme that was started in 1972 in response to widespread concern in Scandinavian countries about the effects of acid precipitation (Overrein *et al.*, 1980; Mohn and Volden, 1985), which was the subject of Example 1.2. Table 1.1 contains the recorded values for acidity (pH) , sulphate (SO_4), nitrate (NO_3), and calcium (Ca) for lakes sampled in 1976, 1977, 1978 and 1981, and Figure 1.2 shows the pH values in the different years plotted against the locations of the lakes.

Consider just the observations for pH and SO_4 for the 46 lakes for which these were recorded in 1976. These are plotted against the location of the lakes in Figure 9.2. In terms of the spatial distribution of the data, some questions that might be asked are:

- Is there any evidence that pH values tend to be similar for lakes that are close in space, i.e., is there any spatial autocorrelation?

- For SO_4, is there any spatial correlation and, if so, is this more or less pronounced than the spatial correlation for pH?

- Is there a significant relationship between the pH and SO_4 measurements, taking into account any patterns that exist in the spatial distributions for each of the measurements considered individually?

In considering these three examples, the most obvious differences between them are that the first concerns counts of the number of shellfish in quadrats, the second concerns the exact location of individual ant nests, and the third concerns the measurement of variables at lakes in particular locations. In addition, it can be noted that the shellfish and lake examples concern sample data in the sense that values could have been recorded at other locations in the study area, but this was not done. By contrast, the positions of all the ant nests in the study area are known, so this is population data.

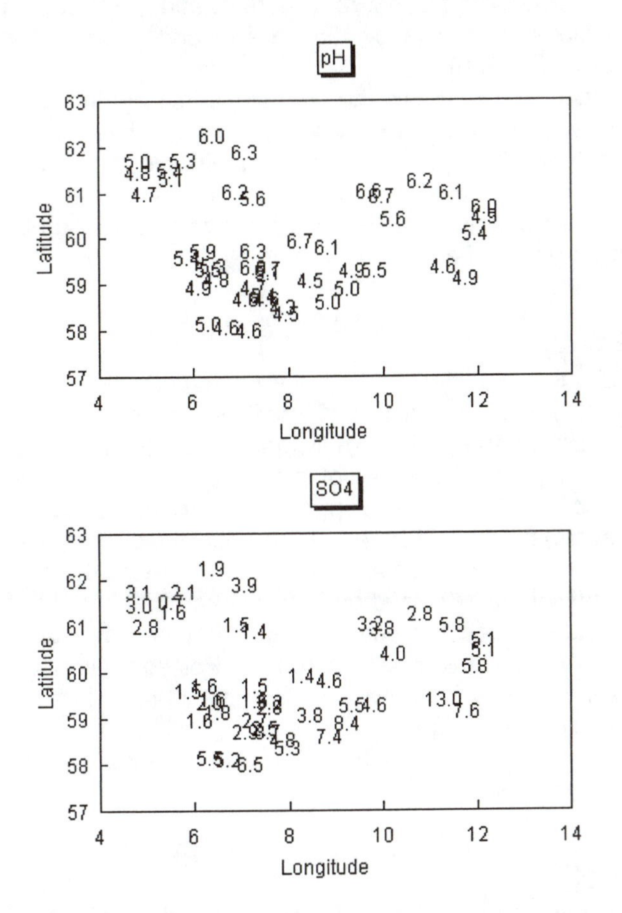

Figure 9.2 Values for pH and SO_4 concentrations (milligrams per litre) of lakes in Norway plotted against the latitude and longitude of the lakes.

9.3 Spatial Patterns in Quadrat Counts

Given a set of quadrat counts over an area, such as those for the pipi in Table 9.1, there may be interest in knowing whether there is any evidence for a spatial pattern. For example, there might be a tendency for the higher counts to be close together (clustering) or to be spread out over the study area (uniformity). The null hypothesis of interest is then randomness in the sense that each of the counts could equally well have occurred in any of the spatial locations. This question has been of particular interest to those studying the spatial distribution of plants where distributions may appear to be random for some quadrat sizes but not for others (Mead, 1974; Galiano *et al.*, 1987; Dale and MacIsaac, 1989; Perry and Hewitt, 1991; Perry, 1995a, 1995b, 1998).

A particular hypothesis that it is sometimes of interest to test is complete spatial randomness, in which case the individual items being counted over a large area are each equally likely to be anywhere in the area, independently of each other. In this case it is a standard result that the counts in quadrats will have a Poisson distribution. That is, the probability of a count x is given by the equation

$$P(x) = \mu^x \exp(-\mu) / x!, \qquad (9.1)$$

where μ is the expected value (mean) of the counts.

One of the characteristics of the Poisson distribution is that the variance equals the mean. Therefore, if $\bar{x}$ and s^2 are the mean and variance, respectively, of a set of quadrat counts, then the variance to mean ratio $R = s^2/\bar{x}$ should be approximately one. Values of R that are much less than one indicate that the counts are more uniform than expected from the Poisson distribution, so that there is some tendency for the individuals to spread out evenly over the study area. On the other hand, values of R much larger than one indicate that the counts are more variable than expected from the Poisson distribution, so that there is a tendency for individuals to accumulate in just a few of the quadrats, i.e., clustering.

A standard test for whether R is significantly different from one involves comparing

$$T = (R - 1)/\sqrt{\{2/(n - 1)\}} \qquad (9.2)$$

with the t-distribution with n - 1 degrees of freedom (df). If a significant result is obtained, then $R < 1$ implies some evenness in the counts, while $R > 1$ implies some clustering in the counts.

Even when the quadrat counts do not follow a Poisson distribution, they can be randomly distributed in space, in the sense that the counts are effectively randomly distributed to the quadrats, independently of each other. In particular, there will be no tendency for the high counts to be close together, so that there is some clustering (positive spatial correlation), or for the high counts to be spread out, so that there is some evenness (negative spatial correlation). This hypothesis can be tested using a Mantel matrix randomization test, which is a type of procedure that was first proposed by Mantel (1967).

If quadrat counts tend to be similar for quadrats that are close in space, then this can be expected to show up in a positive correlation between the spatial distance apart of two quadrats and the absolute difference between the counts in the quadrats. The following test is designed to see whether this correlation is significantly large.

Suppose that there are n quadrats, and let the spatial distance apart of quadrats i and j be denoted by d_{ij}. This distance can be calculated for every pair of quadrats to produce a matrix

$$\mathbf{D} = \begin{bmatrix} 0 & d_{1,2} & d_{1,3} & \cdots & d_{1,n} \\ d_{2,1} & 0 & d_{2,3} & \cdots & d_{2,n} \\ . & . & . & \cdots & . \\ . & . & . & \cdots & . \\ . & . & . & \cdots & . \\ d_{n-1,1} & d_{n-1,2} & d_{n-1,3} & \cdots & d_{n-1,n} \\ d_{n,1} & d_{n,2} & d_{n,3} & \cdots & 0 \end{bmatrix} \qquad (9.3)$$

of geographical distances. Because of the way that it is calculated, this matrix is symmetric with $d_{i,j} = d_{j,i}$, with zeros down the diagonal, as shown.

A second matrix

$$\mathbf{C} = \begin{bmatrix} 0 & c_{1,2} & c_{1,3} & \cdots & c_{1,n} \\ c_{2,1} & 0 & c_{2,3} & \cdots & c_{2,n} \\ . & . & . & \cdots & . \\ . & . & . & \cdots & . \\ . & . & . & \cdots & . \\ c_{n-1,1} & c_{n-1,2} & c_{n-1,3} & \cdots & c_{n-1,n} \\ c_{n,1} & c_{n,22} & c_{n,3} & \cdots & 0 \end{bmatrix} \qquad (9.4)$$

can also be constructed such that the element in row i and column j is the absolute difference

$$c_{i,j} = |c_i - c_j| \qquad (9.5)$$

between the count for quadrat i and the count for quadrat j. Again this matrix is symmetric.

Given these matrices, the question of interest is whether the Pearson correlation coefficient that is observed between the pairs of distances $(d_{2,1},c_{2,1})$, $(d_{3,1},c_{3,1})$, $(d_{3,2},c_{3,2})$, ... $(d_{162,161},c_{162,161})$ in the two matrices is unusually large in comparison with the distribution of this correlation that is obtained if the quadrat counts are equally likely to have had any of the other possible allocations to quadrats. This is tested by comparing the observed correlation with the distribution of correlations that is obtained when the quadrat counts are randomly reallocated to the quadrats.

Mantel (1967) developed this procedure for the problem of determining whether there is contagion with the disease leukaemia, which should show up with cases that are close in space also tending to be close in time. He noted that in practice it may be better to replace spatial distances with their reciprocals and see whether there is a significant negative correlation between $1/d_{i,j}$ and $c_{i,j}$. The reasoning behind this idea is that the pattern that is most likely to be present is a similarity between values that are close together rather than large differences between values that are distant from each other. By its nature, the reciprocal transformation emphasizes small distances and reduces the importance of large differences.

Perry (1995a, 1995b, 1998) discusses other approaches for studying the distribution of quadrat counts. One relatively simple idea is to take the observed counts and consider these to be located at the centres of their quadrats. The individual points are then moved between quadrats, in order to produce a configuration with equal numbers in each quadrat, as close as possible. This is done with the minimum possible amount of movement, which is called the distance to regularity, D. This distance is then compared to the distribution of such distances that is obtained by randomly reallocating the quadrat counts to the quadrats, and repeating the calculation of D for the randomized data. The randomization is repeated many times and the probability, P_a, of obtaining a value of D as large as that observed is calculated. Finally, the mean distance to regularity for the randomized data, E_a, is calculated, and hence the index $I_a = D/E_a$ of clustering. These calculations are done as part of the SADIE (Spatial Analysis by Distance IndicEs) programs that are available from Perry (2000).

Example 9.1 Distribution and Spatial Correlation for Shellfish

For an example, consider the pipi and cockle counts shown in Table 9.1. There are n = 88 counts for both species, and the mean and variance for the pipi counts are $\bar{x} = 19.26$ and $s^2 = 2551.37$. The variance to mean ratio is huge at R = 132.46, which is overwhelmingly significant according to the test of equation (9.2), giving T = 872.01 with 87 df. For cockles the mean and variance are $\bar{x} = 1.24$ and $s^2 = 11.32$, giving R = 9.14 and T = 53.98, again with 87 df. The variance to mean ratio for cockles is still overwhelmingly significant, although it is much smaller than the ratio for pipis. Anyway, for both species the hypothesis that the individuals are randomly located over the study area is decisively rejected. In fact, this is a rather unlikely hypothesis in the first place for something like shellfish.

Although the individuals are clearly not randomly distributed, it could be that the quadrat counts are, in the sense that the observed configuration is effectively a random allocation of the counts to the quadrats without any tendency for the high counts to occur together. Looking at the distribution of pipi counts this seems unlikely, to put it mildly. With the cockles it is perhaps not so obvious that the observed configuration is unlikely. At any rate, for the purpose of this example the Mantel matrix randomization test has been applied for both species.

There are 88 quadrats, and therefore the matrices (9.3) and (9.4) are of size 88 by 88. Of course, the geographical distance matrix is the same for both species. Taking the quadrats row by row from Table 9.1, with the first 11 in the top row, the next 11 in the second row, and so on, this matrix takes the form

$$\mathbf{D} = \begin{bmatrix} 0.0 & 20.0 & 40.0 & \cdots & 200.0 \\ 20.0 & 0.0 & 20.0 & \cdots & 180.0 \\ . & . & . & \cdots & . \\ . & . & . & \cdots & . \\ . & . & . & \cdots & . \\ 193.1 & 174.6 & 156.5 & \cdots & 20.0 \\ 211.9 & 193.1 & 174.6 & \cdots & 0.0 \end{bmatrix},$$

with distances in metres.

For pipis, the matrix of absolute count differences takes the form

$$\mathbf{C} = \begin{bmatrix} 0 & 1 & 3 & \cdots & 29 \\ 1 & 0 & 4 & \cdots & 30 \\ . & . & . & \cdots & . \\ . & . & . & \cdots & . \\ . & . & . & \cdots & . \\ 1 & 0 & 4 & \cdots & 30 \\ 29 & 30 & 26 & \cdots & 0 \end{bmatrix}.$$

The correlation between the elements in the bottom triangular part of this matrix and the elements in the bottom triangular part of the **D** matrix is -0.11, suggesting that as quadrats get further apart, their counts tend to become more similar, which is certainly not what might be expected. When the observed correlation of -0.11 was compared with the distribution of correlations obtained by randomly reallocating the pipi counts to quadrats 10,000 times (Manly, 1997a, Section 5.3) it was found that the percentage of times that a value as far from zero as -0.11 was obtained was only 2.1%. Hence, this negative correlation is significantly different from zero at the 5% level (p = 0.021).

To understand what is happening here, it is useful to look at Figure 9.3(a). This shows that for all except the furthest distances apart the difference between quadrat counts can be anything between 0 and 250. However, for distances greater than about 150m apart, the largest count difference is only about 60. It is this that leads to the negative correlation, which can be seen to be more or less inevitable from the fact that the highest pipi counts are toward the centre of the sampled area. Thus the significant result does seem to be the result of a non-random distribution of counts, but this non-randomness does not show up as a simple tendency for the counts to become more different as they get further apart.

For cockles, there is little indication of any correlation either positive or negative from a plot of the absolute count differences against their distances apart that is shown in Figure 9.3(b). The correlation is 0.07, which is not significant by the Mantel randomization test (p = 0.15).

For these data the use of reciprocal distances as suggested by Mantel (1967) is not helpful. The correlation between pipi count differences and the reciprocal of the distances apart of quadrats is 0.03, which is not at all significant by the randomization test (p = 0.11), and the correlation between cockle count differences and reciprocal distances is -0.03, which is also not significant (p = 0.14). Plots of the count differences against the reciprocal distances are shown in Figure 9.3 (b) and (d).

There are other variations on the Mantel test that can be applied with the shellfish data. For example, instead of using the quadrat counts, these can just be coded to 0 (absence of shellfish) and 1 (presence of shellfish), which then avoids the situation where two quadrats have a large difference in counts, but both also have a large count relative to most quadrats. A matrix of presence-absence differences between quadrats can then be calculated, where the ith row and jth column contain 0 if quadrat i and quadrat j either both have a shellfish species, or are both missing it, and 1 if there is one presence and one absence. If this is done, then it turns out that there is a significant negative correlation of -0.047 between the presence-absence differences for pipi and the reciprocal distances between quadrats (p = 0.023). Other correlations (presence-absence differences for pipi versus distances, and presence-absence differences for cockles versus distances and reciprocal distances) are close to zero and not at all significant.

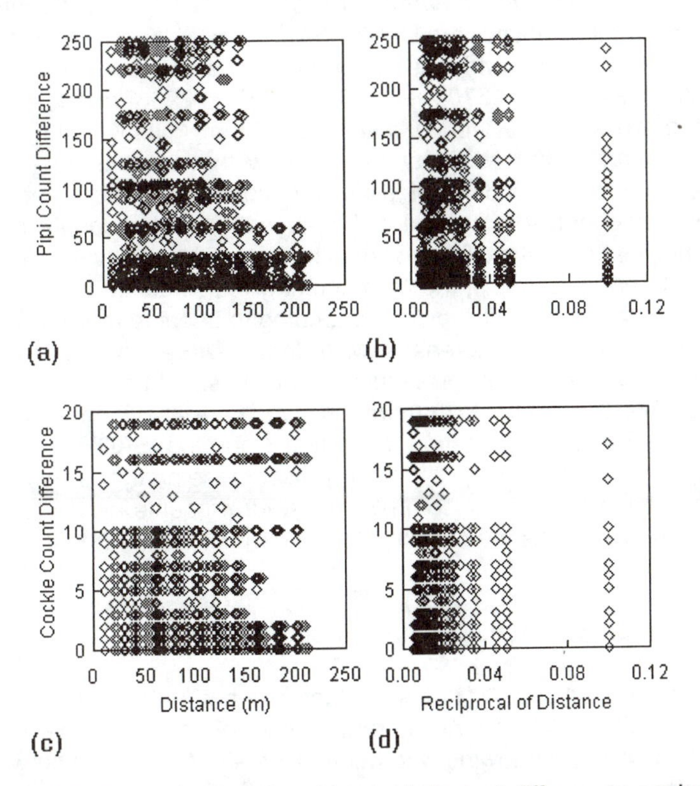

Figure 9.3 Plots of pipi and cockle quadrat count differences against the differences apart of the quadrats, and the reciprocals of the distances apart.

One problem with using the Mantel randomization test is that it is not usually an option in standard statistical packages, so that a special computer program is needed. The calculations described for this example, including the construction of distance matrices, were carried out using the computer package RT for randomization testing (Manly, 1997b), so the use of this is one possibility. Other programs such as

MANTEL-STRUCT can be downloaded from the internet (Miller, 1998).

The SADIE approach was also applied to the pipi and cockle data, using a Windows program called RBRELV13.EXE (Perry, 2000). Included in the output from the program are the statistics D (the distance to regularity), P_a (the probability of a value for D as large as that observed occurring for randomly allocated quadrat counts), E_a (the mean distance to regularity for randomly allocated counts), and $I_a = D/E_a$ (the index of non-randomness in the distribution of counts). For pipis these statistics are D = 48963, P_a = 0.26, E_a = 44509, and I_a = 1.10. There is no evidence here of non-randomness. For cockles the statistics are D = 5704, P_a = 0.0037, E_a = 2968, and I_a = 1.92. Here, there is clear evidence of some clustering.

In summary, the above analyses show that there is very clear evidence that the individual pipis and cockles are not randomly distributed over the sampled area because the quadrat counts for the two species do not have Poisson distributions. Given the values of the quadrat counts for pipis, the Mantel test gives evidence of structure that is not as simple as the counts tending to become more different as the distance apart increases, while the SADIE analysis based on the distance to regularity gives no evidence of structure. By contrast, for cockles, the Mantel test approach gives no real evidence of non-randomness in the location of counts, but the SADIE analysis indicates some clustering. If nothing else, these analyses show that the evidence for non-randomness in quadrat counts depends very much on how this is measured.

9.4 Correlation Between Quadrat Counts

Suppose that two sets of quadrat counts are available, such as those shown in Table 9.1 for two different species of shellfish. There may then be interest in knowing whether there is any association between the sets of counts in the sense that high counts tend to occur either in the same locations (positive correlation) or in different locations (negative correlation).

Unfortunately, testing for an association in this context is not a straightforward matter. Besag and Diggle (1977) and Besag (1978) proposed a randomization test that is a generalization of Mead's (1974) test for randomness in the spatial distribution of a single species which can be used when the number of quadrats is a multiple of 16. However, this test is not valid if there is spatial correlation within each of the sets of counts being considered (Manly, 1997a, p. 215).

At present the most promising method for handling the problem seems to be along the lines of one that is proposed by Perry (1998). This is based on an algorithm for producing permutations of quadrat counts with a fixed level of aggregation and a fixed centroid over the study region, combined with a test statistic based on the minimum amount of movement between quadrats that is required in order to produce counts that are all equal. One test statistic is derived as follows:

(a) The counts for the first variable are scaled by multiplying by the sum of the counts (M_2) for the second variable, and the counts for the second variable are scaled by multiplying by the sum of the counts (M_1) for the first variable. This has the effect of making the scaled total count equal to M_1M_2 for both variables.

(b) The scaled counts for both variables are added for the quadrats being considered.

(c) A statistic T is calculated, where this is the amount of movements of individuals between quadrats that is required to produce an equal number of scaled individuals in each quadrat.

(d) The statistic T is compared with the distribution of the same statistic that is obtained by permuting the original counts for variable 1, with the permutations constrained so that the spatial distribution of the counts is similar to that of the original data. An index of association, $I_{t(2)1} = T/E_{t(2)1}$, is also computed, where $E_{t(2)1}$ is the mean of T for the permuted sets of data. The significance level of $I_{t(2)1}$ is the proportion of permutations that give a value as larger or larger than the observed value.

(e) The observed statistic T is also compared with the distribution that is obtained by permuting the original counts for variable 2, again with constraints to ensure that the spatial distribution of the counts is similar to that for the original data. Another index of association is then $I_{t(1)2} = T/E_{t(1)2}$, where $E_{t(1)2}$ is the mean number of steps to regularity for the permuted sets of data. The significance level of $I_{t(1)2}$ is the proportion of permutations that give a value as large or larger than the observed value.

The idea with this approach is to compare the value of T for the real data with the distribution that is obtained for alternative sets of data for which the spatial distribution of quadrat counts is similar to that for the real data for each of the two variables being considered.

This is done by constraining the permutations for the quadrat counts so that the distance to regularity is close to that for the real data, as is also the distance between the centroid of the counts and the centre of the study region. A special algorithm is used to find the minimum number of steps to regularity. Perry (1998) should be consulted for more details about how this is achieved.

If the two variables being considered tend to have large counts in the same quadrats, then T will be large, with any clustering in the individual variables being exaggerated by adding the scaled counts for the two variables. Hence values of $I_{t(2)1}$ and $I_{t(1)2}$ greater than one indicate an association between the two sets of counts. On the other hand, values of $I_{t(2)1}$ and $I_{t(1)2}$ of less than one indicate a tendency for the highest counts to be in different quadrats for the two variables.

Apparently, in practice $I_{t(2)1}$ and $I_{t(1)2}$ tend to be similar, as do their significance levels. In that case, Perry (1998) suggests averaging $I_{t(2)1}$ and $I_{t(1)2}$ to get a single index I_t, and averaging the two significance levels to get a combined significance level P_t.

Perry (1998) suggests other test statistics that can be used to examine the association between quadrat counts, and is still developing the SADIE approach for analysing quadrat counts. See Perry (2000) for more about these methods.

Example 9.2 Correlation Between Counts for Pipis and Cockles

To illustrate the SADIE approach for assessing the association between two sets of quadrat counts, consider again the pipi and cockle counts given in Table 9.1. It may be recalled from Example 9.1 that these counts clearly do not have Poisson distributions, so that the individual shellfish are not randomly located, and there is mixed evidence concerning whether the counts themselves are randomly located. Now the question to be considered is whether the pipi and cockle counts are associated in terms of their distributions.

The data were analysed with the computer program SADIEA (Perry, 2000). The distance to regularity for the total of the scaled quadrat counts is T = 129196 m. When 1000 randomized sets of data were produced keeping the cockle counts fixed and permuting the pipi counts, the average distance to regularity was $E_{t(2)1}$ = 105280 m, giving the index of association $I_{t(2)1}$ = 129196/105280 = 1.23. This was exceeded by 33% of the permuted sets of data, giving no real evidence of association. On the other hand, keeping the pipi counts fixed and permuting the cockle counts gave an average distance to regularity of $E_{t(1)2}$ = 100062, giving the index of association $I_{t(1)2}$ =

129196/100062 = 1.29. This was exceeded by 3.4% of the permuted sets of data. In this case there is some evidence of association.

The conclusion from the two randomizations are quite different in terms of the significance of the results, although both $I_{t(2)1}$ and $I_{t(1)2}$ suggest some positive association between the two types of count. The reason for the difference in terms of significance is presumably the result of a lack of symmetry in the relationship between the two shellfish. In every quadrat where cockles are present, pipis are present as well. However, cockles are missing from half of the quadrats where pipis are present. Therefore, apparently cockles are associated with pipis, but pipis are not necessarily associated with cockles.

9.5 Randomness of Point Patterns

Testing whether a set of points appears to be randomly located in a study region is another problem that can be handled by computer-intensive methods. One approach was suggested by Besag and Diggle (1977). The basic idea is to calculate the distance from each point to its nearest neighbour and calculate the mean of these distances. This is then compared with the distribution of the mean nearest neighbour distance that is obtained when the same number of points are allocated to random positions in the study area.

An extension of this approach uses the mean distances between second, third, fourth, etc. nearest neighbours as well as the mean of the first nearest neighbour distances. Thus let Q_i denote the mean distance between each point and the point that is its ith nearest neighbour. For example, Q_3 is the mean of the distances from points to the points that are third closest to them. The observed configuration of points then yields the statistics Q_1, Q_2, Q_3, and so on. These are compared with the distributions for these mean values that are generated by randomly placing the same number of points in the study area, with a computer simulation being carried out to produce these distributions.

This type of procedure is sometimes called a Monte Carlo test. With these types of tests there is freedom in the choice of the test statistics to be employed. They do not have to be based on nearest neighbour distances and, in particular, the use of Ripley's (1981) K-function is becoming popular (Andersen, 1992; Haase, 1995).

Example 9.3 The Location of Messor Wasmanni *Nests*

As an example of the Monte Carlo test based on nearest neighbour distances that has just been described, consider the locations of the 45 nests of *Messor wasmanni* that are shown in Figure 9.1. The values for the mean nearest neighbour distances Q_1 to Q_{10} are shown in the second column of Table 9.2, and the question to be considered is whether these values are what might reasonably be expected if the 45 nests are randomly located over the study area.

Table 9.2 Results from testing for randomness in the location of *Messor wasmanni* nests using nearest neighbour statistics

i	Observed value of Q_i	Simulation mean of Q_i	Percentage significance level Lower	Upper
1	22.95	18.44	99.82	0.20
2	32.39	28.27	99.14	0.88
3	42.04	35.99	99.94	0.08
4	50.10	42.62	99.96	0.06
5	56.58	48.60	99.98	0.04
6	61.03	54.13	99.56	0.46
7	65.60	59.34	98.30	1.72
8	70.33	64.26	97.02	3.00
9	74.45	68.99	94.32	5.70
10	79.52	73.52	94.78	5.24

The Monte Carlo test proceeds as follows:

(a) A set of data for which the null hypothesis of spatial randomness is true is generated by placing 45 points in the study region in such a way that each point is equally likely to be anywhere in the region.

(b) The statistics Q_1 to Q_{10} are calculated for this simulated set of data.

(c) Steps (a) and (b) are repeated 5000 times.

(d) The lower tail significance level for the observed value of Q_i is calculated as the percentage of times that the simulated values for Q_i are less than or equal to the observed value of Q_i.

(e) The upper tail significance level for the observed value of Q_i is calculated as the percentage of times that the simulated values of Q_i are greater than or equal to the observed value of Q_i.

Non-randomness in the distribution of the ant nests is expected to show up as small percentages for the lower or upper tail significance levels, which indicates that the observed data were unlikely to have arisen if the null hypothesis of spatial randomness is true.

The last two columns of Table 9.2 show the lower and upper tail significance levels that were obtained from this procedure when it was carried out using an option in the computer package RT (Manly, 1997b). It is apparent that all of the observed mean nearest neighbour distances are somewhat large, and that the mean distances from nests to their six nearest neighbours are much larger than what is likely to occur by chance with spatial randomness. Instead, it seems that there is some tendency for the ant nests to be spaced out over the study region. Thus the observed configuration does not appear to be the result of the nests being randomly located.

9.6 Correlation Between Point Patterns

Monte Carlo tests are also appropriate for examining whether two point patterns seem to be related, as might be of interest, for example, with the comparison of the positions of the nests for the *Messor wasmanni* and *Cataglyphis bicolor* ants that are shown in Figure 9.1. In this context Lotwick and Silverman (1982) suggested that a point pattern in a rectangular region can be converted to a pattern over a larger area by simply copying the pattern from the original region to similar sized regions above, below, to the left, and to the right, and then copying the copies as far away from the original region as required. A test for independence between two patterns then involves comparing a test statistic observed for the points over the original region with the distribution of this statistic that is obtained when the rectangular 'window' for the species 1 positions is randomly shifted over an enlarged region for the species 2 positions. Harkness and Isham (1983) used this type of analysis with the ant data and concluded that there is evidence of a relationship between the positions of the two types of nests. See also Andersen's (1992) review of these types of analyses in ecology.

As Lotwick and Silverman note, the need to reproduce one of the point patterns over the edge of the region studied in an artificial way is an unfortunate aspect of this procedure. It can be avoided by taking the rectangular window for the type 1 points to be smaller than the total area covered and calculating a test statistic over this smaller area. The distribution of the test statistic can then be determined by randomly placing this small window within the larger area a large number of times. In this case the positions of any type 1 points outside the small window are ignored and the choice of the positioning of the small window within the larger region is arbitrary.

Another idea involves considering a circular region and arguing that if two point patterns within the region are independent, then this means that they have a random orientation with respect to each other. Therefore a distribution that can be used to assess a test statistic is the one that is obtained from randomly rotating one of the sets of points about the centre point of the region. A considerable merit with this idea is that the distribution can be determined as accurately as desired by rotating one of the sets of points about the centre of the study area from zero to 360 degrees in suitable small increments (Manly, 1997a, Section 9.7).

The method of Perry (1998) that has been described in Section 9.4 can be applied with a point pattern as well as with quadrat counts. From an analysis based on this method Perry (1998) concluded that there is no evidence of association between the positions of the nests for the two species of ants that are shown in Figure 9.1.

9.7 Mantel Tests for Autocorrelation

With a variable measured at a number of different positions in space such as the values of pH and SO_4 that are shown in Figure 9.3, one of the main interests is often to test for significant spatial autocorrelation, and to characterize this correlation if it is present. The methods that can be used in this situation are extensive, and no attempt will be made here to review them in any detail. Instead, a few simple analyses will be briefly described.

If spatial autocorrelation is present, then it will usually be the case that this is positive, so that there is a tendency for observations that are close in space to have similar values. Such a relationship is conveniently summarised by the plotting some measure of the difference between two observations against some measure of the spatial difference apart of the observations, for every possible pair of observations. The significance of the autocorrelation can then be tested using the Mantel (1967) randomization test that has been described in Section 9.3.

Example 9.4 Autocorrelation in Norwegian Lakes

Figure 9.4 shows four plots constructed from the pH data of Figure 9.2. Part (a) of the figure shows the absolute differences in pH values plotted against the geographical distances apart for the 1035 possible pairs of lakes. Here the smaller pH differences tend to occur with both the smallest and largest geographical distances, and the correlation between the pH differences and the geographical distances of r = 0.034 is not significantly large (p = 0.276, from a Mantel test with 5000 randomizations). Part (b) of the figure shows the absolute pH differences plotted against the reciprocals of the geographical distances. The correlation here is -0.123, which has the expected negative sign, and is highly significantly low (p = 0.0008, with 5000 randomizations). Part (c) of the figure shows 0.5(pH difference)^2 plotted against the geographical distance. The reason for considering this plot is that it corresponds to a variogram cloud, in the terminology of geostatistics, as discussed further below. The correlation here is r = 0.034, which has the correct sign but is not significant (p = 0.286, with 5000 randomizations). Finally, part (d) of the plot shows the 0.5(pH difference)^2 values plotted against the reciprocals of geographical distances. Here the correlation of r = -0.121 has the correct sign and is highly significantly low (p = 0.0006, with 5000 randomizations).

The significant negative correlations for the plots in parts (b) and (d) of Figure 9.4 give clear evidence of spatial autocorrelation, with close lakes tending to have similar pH values. It is interesting to see that this evidence appears when the measures of pH difference are plotted against reciprocals of geographical distances rather than plotted against the geographical distances themselves, as is commonly done with geostatistical types of analysis.

By contrast, when plots and correlations are produced for the SO_4 variable for which the data are shown in the lower part of Figure 9.2, then spatial autocorrelation is more evident when the measures of SO_4 differences are plotted against the geographical distances rather than their reciprocals (Figure 9.5). However, the spatial autocorrelation is altogether stronger for SO_4 than it is for pH, and is highly significant for all four types of plots:

(a) for absolute SO_4 differences plotted against geographical distances, r = 0.30, p = 0.002;

(b) for absolute SO_4 differences plotted against reciprocal geographical distances, r = -0.23, p = 0.002;

(c) for half-squared SO_4 differences plotted against geographical distances, $r = 0.25$, $p = 0.0008$; and

(d) for half-squared SO4 differences plotted against reciprocal geographical distances, $r = -0.16$, $p = 0.012$.

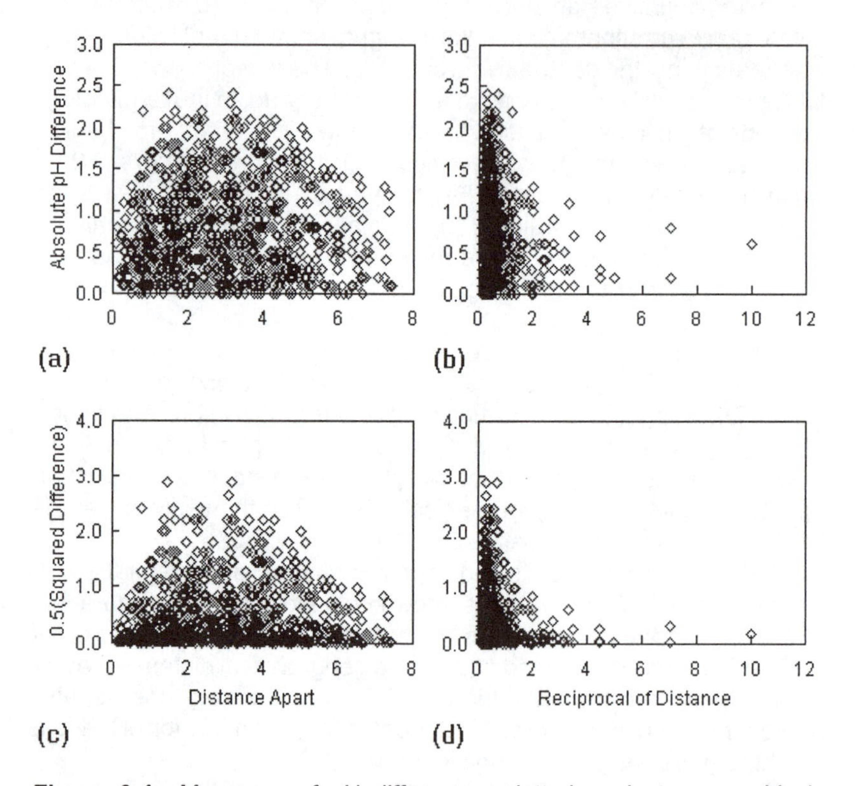

Figure 9.4 Measures of pH differences plotted against geographical distances (measured in degrees of latitude and longitude) and reciprocals of geographical distances for pairs of Norwegian lakes.

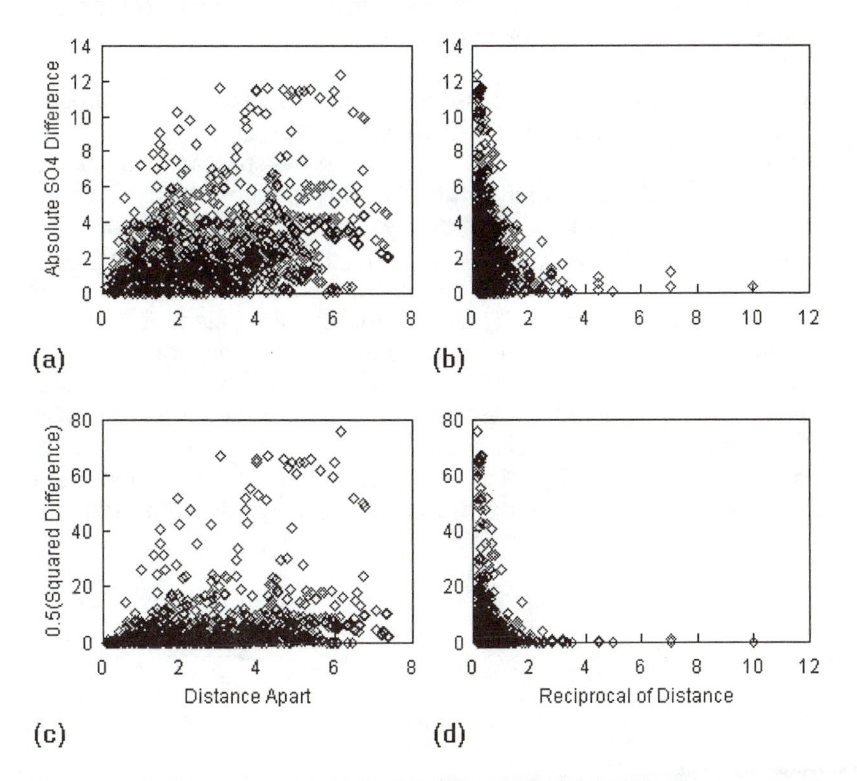

Figure 9.5 Measures of SO_4 differences plotted against geographical distances (measured in degrees of latitude and longitude) and reciprocals of geographical distances for pairs of Norwegian lakes.

9.8 The Variogram

Geostatistics is the name given to a range of methods for spatial data analysis that were originally developed by mining engineers for the estimation of the mineral resources in a region, based on the values measured at a sample of locations (Krige, 1966; David, 1977; Journel and Huijbregts, 1978). A characteristic of these methods is that an important part is played by a function called the variogram (or sometimes the semivariogram), which quantifies the extent to which values tend to become more different as pairs of observations become further apart, assuming that this does in fact occur.

To understand better what is involved, consider part (c) of Figure 9.5. This plot was constructed from the SO_4 data for the 46 lakes with the positions shown in Figure 9.2. There are 1035 pairs of lakes, and each pair gives one of the plotted points. What is plotted on the vertical axis for the pair consisting of lake i with lake j is

$$D_{ij} = 0.5(y_i - y_j)^2,$$

where y_i is the SO_4 level for lake i. This is plotted against the geographical distance apart of the lakes on the horizontal axis. There does indeed appear to be a tendency for D_{ij} to increase as lakes get further apart, and a variogram can be used to quantify this tendency.

A plot like that in Figure 9.5(c) is sometimes called a variogram cloud. The variogram itself is a curve through the data that gives the mean value of D_{ij} as a function of the distance apart of the lakes. There are two varieties of this. An experimental or empirical variogram is obtained by smoothing the data to some extent to highlight the trend, as described in the following example. A model variogram is obtained by fitting a suitable mathematical function to the data, with a number of standard functions being used for this purpose.

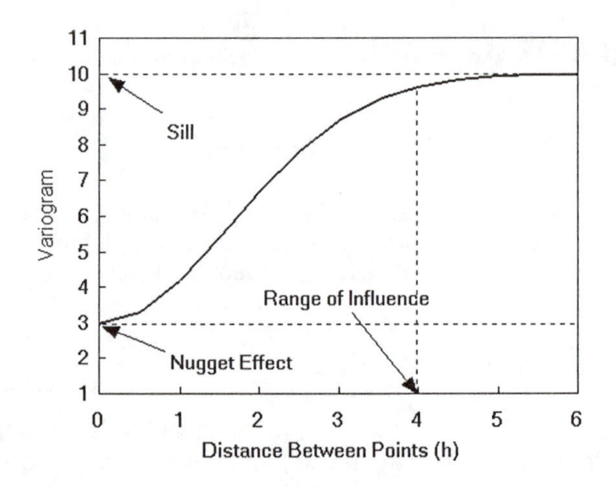

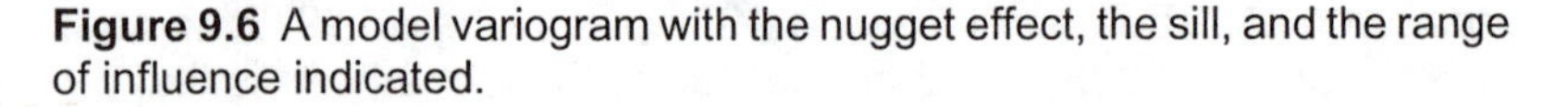

Figure 9.6 A model variogram with the nugget effect, the sill, and the range of influence indicated.

Typically, a model variogram looks something like the one in Figure 9.6. Even two points that are very close together may tend to have different values, so there is a 'nugget effect', with the expected value of $0.5(y_i - y_j)^2$ being greater than zero even with h = 0. In the figure this nugget effect is 3. The maximum height of the curve is called the 'sill'.

In the figure this is 10. This is the maximum value of $0.5(y_i - y_j)^2$, which applies for two points that are far apart in the study area. Finally, the 'range of influence' is the distance apart at which two points have effectively independent values. This is sometimes defined as the point at which the curve is 95% of the difference between the nugget and the sill. In the figure the range of influence is 4.

There are a number of standard mathematical models for variograms. One is the Gaussian model, with the equation

$$\gamma(h) = c + (S - c)\{1 - \exp(-3h^2/a^2)\}. \tag{9.6}$$

Here c is the nugget effect, S is the sill, and a is the range of influence. When h = 0 the exponential term equals 1, so that $\gamma(0) = c$. When h is very large the exponential term becomes insignificant, so that $\gamma(\infty) = S$. When h = a, the exponential term becomes exp(3) = 0.050, so that $\gamma(a) = c + 0.95(S - c)$.

Other models that are often considered are the spherical model with the equation

$$\gamma(h) = \begin{cases} c + (S - c)\{1.5(h/a) - 0.5(h/a)^3, & h \leq a \\ c, \text{ otherwise,} & \end{cases} \tag{9.7}$$

the exponential model with the equation

$$\gamma(h) = c + (S - c)\{1 - \exp(-3h/a)\}, \tag{9.8}$$

and the power model with

$$\gamma(h) = c + Ah^w. \tag{9.9}$$

For all of these models, c is the nugget effect. The spherical and exponential models also have a sill at S, but for the power model the function increases without limit as h increases. For the spherical model the sill is reached when h = a, for the exponential model a is an effective range of influence as $\gamma(a) = c + 0.95(S - c)$, and for the power model the range of influence is infinite.

A variogram describes the nature of spatial correlation. It may also give information about the correlation between points that are separated by different spatial distances. This is because if Y_i and Y_j are values of a random variable measured at two locations with the same mean μ, then the expected value of half of the difference squared is

$$E\{0.5(Y_i - Y_j)^2\} = 0.5\{(Y_i - \mu)^2 - 2(Y_i - \mu)(Y_j - \mu) + (Y_j - \mu)^2\}$$
$$= 0.5\{Var(Y_i) - 2Cov(Y_i,Y_j) + Var(Y_j)\}.$$

Hence if the variance is equal to σ^2 at both locations, it follows that

$$E\{0.5(Y_i - Y_j)^2\} = \sigma^2 - Cov(Y_i,Y_j).$$

As the correlation between Y_i and Y_j is $\rho(Y_i,Y_j) = Cov(Y_i,Y_j)/\sigma^2$ under these conditions, it follows that

$$E\{0.5(Y_i - Y_j)^2\} = \sigma^2\{1 - \rho(Y_i,Y_j)\}.$$

Suppose also that the correlation between Y_i and Y_j is only a function of their distance apart, h. This correlation can then be denoted by $\rho(h)$. Finally, note that the left-hand side of the equation is actually the variogram for points distance h apart, so that

$$\gamma(h) = \sigma^2\{1 - \rho(h)\}. \quad (9.10)$$

This is therefore the function that the variogram is supposed to describe in terms of the variance of the variable being considered and the correlation between points at different distances apart. One important fact that follows is that because $\rho(h)$ will generally be close to zero for large values of h, the sill of the variogram should equal σ^2, the variance of the variable Y being considered.

The requirements for equation (9.10) to hold are that the mean and variance of Y are constant over the study area, and that the correlation between Y at two points depends only on their distance apart. This is called second order stationarity. Actually, the requirement for the variogram to exist and to be useful for data analysis is less stringent. This is the so-called intrinsic hypothesis, that the mean of the variable being considered is the same at all locations, with the expected value of $0.5(Y_i - Y_j)^2$ depending only on the distance between the locations of the two points (Pannatier, 1996, Appendix A).

Example 9.5 Variograms for SO4 Values of Norwegian Lakes

Figure 9.7 shows experimental and model variograms estimated for the SO_4 data from Norwegian lakes that are shown in Figure 9.2. The experimental variogram in this case was estimated by taking the maximum distance between two lakes and dividing this into 12 class intervals centred at 0.13, 0.39, 0.73, ..., 3.85. The values for $0.5(y_i - y_j)^2$ were then averaged within each of these intervals to produce the points that are plotted. For example, for the interval centred at 0.13, covering pairs of points with a distance between them in the range from 0.00 to 0.26, the mean value of $0.5(y_i - y_j)^2$ is 0.29. An equation for this variogram is therefore

$$\hat{\gamma}(h) = \sum_{i=1}^{N(h)} 0.5\,(y_i - y_j)^2 / N(h)\}, \qquad (9.11)$$

where $\hat{\gamma}(h)$ is the empirical variogram value for the interval centred at a distance h between points, N(h) is the number of pairs of points with a distance between them in this interval, and the summation is over these pairs of points.

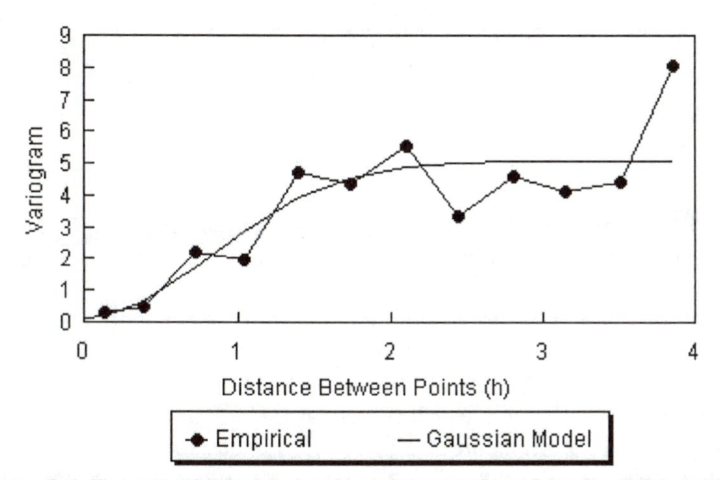

Figure 9.7 Experimental and model variograms found for the SO4 data on Norwegian lakes displayed in Figure 9.2.

The model variogram for Figure 9.7 is the Gaussian model given by equation (9.6). This was estimated using a program produced by the United States Environmental Protection Agency called GEOPACK (Yates and Yates, 1990) to give

$$\gamma(h) = 0.126 + 4.957\{1 - \exp(-3h^2/2.019^2)\}. \qquad (9.12)$$

This is therefore the model function that is plotted on Figure 9.7. The nugget effect is estimated to be 0.126, the sill is estimated to be 0.126 + 4.957 = 5.083, and the range of influence is estimated to be 2.019. Essentially, this model variogram is a smoothed version of the experimental variogram, which is itself a smoothed version of the variogram cloud in Figure 9.5(c).

9.9 Kriging

Apart from being a device for summarising the spatial correlation in data, the variogram is also used to characterize this correlation for many other types of geostatistical analyses. There is a great deal that could be said in this respect. However, only one of the commonest of these analyses will be considered here. This is kriging, which is a type of interpolation procedure named after the mining engineer D.G. Krige, who pioneered these methods.

Suppose that in the study area sample values $y_1, y_2, ..., y_n$ are known at n locations, and it is desired to estimate the value y_0 at another location. For simplicity assume that there are no underlying trends in the values of Y. Then kriging estimates y_0 by a linear combination of the known values,

$$\hat{y}_0 = \sum a_i y_i, \tag{9.13}$$

with the weights $a_1, a_2, ..., a_n$ for these known values chosen so that the estimator of y_0 is unbiased, with the minimum possible variance for prediction errors.

The equations for determining the weights to be used in equation (9.13) are somewhat complicated. They are derived and explained by Thompson (1992, Chapter 20), and Goovaerts (1997, Chapter 5), among others. They are a function of the assumed model for the variogram. To complicate matters, there are also different types of kriging, with resulting modifications to the basic procedure. For example, simple kriging assumes that the expected value of the measured variable is constant and known over the entire study region, ordinary kriging allows for the mean to vary in different parts of the study region by only using close observations to estimate an unknown value, and kriging with trend assumes a smooth trend in the mean over the study area.

Ordinary kriging seems to be what is most commonly used. In practice this is done in three stages:

- The experimental variogram is calculated to describe the spatial structure in the data.

- Several variogram models are fitted to the experimental variogram, either 'by eye' or by non-linear regression methods, and one model is chosen to be the most appropriate.

- The kriging equations are used to produce estimates of the variable of interest at a number of locations that have not been sampled. Often this is at a grid of points covering the study area.

Example 9.6 Kriging with the SO_4 Data

The data on SO_4 from Norwegian lakes provided in Figure 9.2 can be used to illustrate the type of results that can be obtained by kriging. A Gaussian variogram for these data is provided by equation (9.12) and displayed graphically in Figure 9.7. This variogram was used to estimate the SO_4 level for a grid of points over the study region, with the results obtained being shown in Table 9.3 and Figure 9.8. The calculations were carried out using GEOPACK (Yates and Yates, 1990), which produced the estimates and standard errors shown in Table 9.3 using the default values for all parameters. The three-dimensional plot shown in Figure 9.8 was produced using the statistical package NCSS 2000 (Hintze, 1998) using the output saved from GEOPACK.

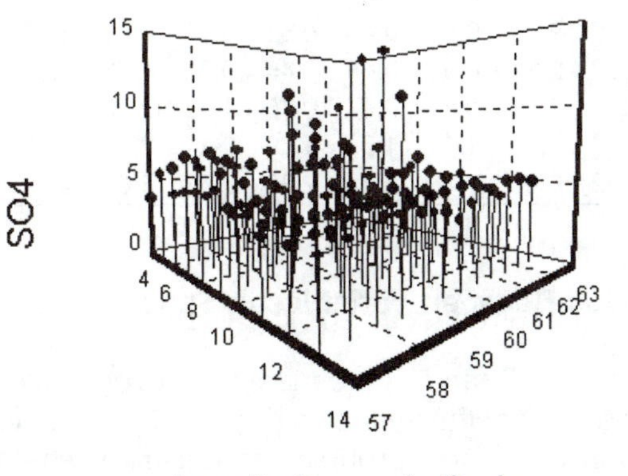

Figure 9.8 Three-dimensional plot of SO_4 levels as estimated by applying ordinary kriging to the data shown in Figure 9.2, using the Gaussian variogram shown in Figure 9.7.

Table 9.3 Estimated SO_4 values obtained by ordinary kriging using the data shown in the lower part of Figure 9.2, with the standard errors associated with these estimated values.

Latitude (° North)	Longitude (°East) 4:00	5:00	6:00	7:00	8:00	9:00	10:00	11:00	12:00	13:00
	Estimate of SO_4 Concentrations (mg/l)									
62:30	4.93	3.31	1.53	2.57	3.22	3.50	2.63	3.15	3.93	5.10
62.00	5.29	2.93	1.80	3.56	3.78	2.82	2.48	2.23	4.57	5.37
61:30	5.37	2.58	1.75	3.09	3.20	2.16	2.62	2.93	5.10	5.36
61:00	5.12	2.80	1.54	1.59	1.39	2.28	3.39	4.23	5.78	5.03
60:30	3.14	2.95	1.76	0.79	0.52	3.76	3.51	6.80	5.68	5.36
60:00	2.36	1.96	1.48	0.80	0.79	4.24	3.69	10.32	6.09	5.52
59:30	2.17	2.14	1.58	1.12	2.48	4.99	4.02	12.66	6.47	5.37
59:00	1.50	2.03	1.57	1.95	3.89	6.75	6.70	12.03	6.22	5.58
58:30	3.74	2.70	3.23	3.67	4.89	8.12	8.70	9.46	6.17	6.43
58:00	3.74	3.70	5.50	6.08	6.41	7.14	9.37	7.98	7.72	7.60
57:30	3.74	4.13	6.46	7.03	6.66	5.57	5.77	9.97	7.60	7.60
57:00	3.74	5.50	6.11	6.81	6.27	6.81	7.40	3.74	3.74	3.74
	Standard Errors of Estimates									
62:30	2.41	1.71	1.16	1.23	2.36	2.58	2.53	2.44	2.65	3.10
62.00	2.07	0.81	0.55	0.51	1.90	2.21	1.99	1.70	2.29	2.80
61:30	1.93	0.43	0.58	0.63	1.68	1.71	1.15	0.78	1.55	2.46
61:00	2.01	0.00	0.83	0.46	1.35	1.35	0.48	0.64	0.64	2.08
60:30	2.45	1.01	1.04	0.73	1.01	1.11	0.53	1.03	0.49	1.81
60:00	2.51	1.59	0.72	0.73	0.62	0.62	0.68	0.96	0.51	2.00
59:30	2.67	1.80	0.44	0.54	0.53	0.49	0.61	0.68	0.71	2.22
59:00	3.10	2.02	0.51	0.48	0.51	0.47	0.85	1.10	0.70	2.36
58:30	2.32	2.29	0.64	0.48	0.45	0.55	1.53	1.93	1.52	2.63
58:00	2.32	2.44	0.81	0.47	0.75	1.35	2.16	2.40	2.49	2.96
57:30	2.32	2.73	1.50	1.21	1.59	2.11	2.67	2.96	2.95	3.10
57:00	2.32	3.00	2.38	2.11	2.33	2.67	3.09	2.32	2.32	2.32

9.10 Correlation Between Variables in Space

When two variables Z_1 and Z_2 are measured over space at the same n locations and each of the variables displays spatial autocorrelation, then this complicates the problem of deciding whether the two variables are correlated with each other. Certainly it is not valid to calculate the correlation using the set of paired data values and treat this as if it is a random sample of independent pairs of observations for the purpose of deciding whether or not the correlation is statistically significant. Instead, some allowance must be made for the effects of the autocorrelation in the individual variables.

In the past, one approach that has been used for handling this problem has used an extension of the Mantel (1967) randomization test as described above. This approach involves calculating three distance matrices. The first is an n by n matrix **A** in which the element a_{ij} in row i and column j is some measure of the difference between the values for Z_1 at locations i and j. The second matrix **B** is of the same size but the element b_{ij} in row i and column j is some measure of the difference between the values of Z_2 for locations i and j. Finally, the third matrix **G** is also of the same size, with the element g_{ij} in row i and column j being the geographical distance between locations i and j. The equation

$$a_{ij} = b_0 + b_1 b_{ij} + b_2 g_{ij}$$

is then fitted to the distances by multiple regression, and the significance of the coefficient b_1 is tested by comparison with the distribution for this coefficient that is obtained when the labels of the n locations are randomly permuted for the matrix **A**.

Unfortunately, the general validity of this procedure has proved to be questionable, although it is possible that it can be applied under certain conditions if the n locations are divided into blocks and a form of restricted randomization is carried out (Manly, 1997a, Section 9.6). It seems fair to say that this type of approach is still promising, but more work is required to better understand how to overcome problems with its use.

An alternative approach is based on geostatistical methods (Liebhold and Sharov, 1997). This involves comparing the observed correlation between two variables with the distribution obtained from simulated sets of data that are generated in such a way that the variogram for each of the two variables matches the one estimated from the real data, but the two variables are distributed independently of each other. The generation of the values at each location is carried out using a technique called sequential unconditional Gaussian simulation (Borgman *et al.*, 1984; Deutsch and Journel, 1992). This method for testing observed correlations is very computer intensive but at present seems to be the best solution to the problem.

9.11 Chapter Summary

- The types of data considered in the chapter are quadrat counts, situations where the location of objects is mapped, and situations where a variable is measured at a number of locations in a study area.

- With quadrat counts, the distribution of the counts will follow a Poisson distribution if objects are randomly located over the study area, independently of each other. The ratio of the variance to the mean (R) is exactly one for a Poisson distribution, and this ratio is often used as a measure of clustering, with $R > 1$ indicating clustering and $R < 1$ indicating regularity in the location of individuals. A t-test is available to decide whether R is significantly different from one.

- The Mantel matrix randomization test can be used to see whether quadrats that are close together tend to have different counts. Other randomization tests for spatial correlation are available with the SADIE (Spatial Analysis by Distance IndicEs) computer programs.

- Testing whether two sets of quadrat counts (e.g., individuals of two species counted in the same quadrats) are associated is not straightforward. A randomization test approach involving producing randomized sets of data with similar spatial structure to the true data appears to be the best approach. One such test is described that is part of the SADIE programs.

- Randomness of point patterns (i.e., whether the points seem to be randomly located in the study area) can be based on comparing the observed mean distances between points and their nearest neighbours with the distributions that are generated by placing the same number of points randomly in the study area using a computer simulation.

- Correlation between two point patterns can also be tested by comparing observed data with computer-generated data. One method involves comparing a statistic that measures the association observed for the real data with the distribution of the same statistic obtained by randomly shifting one of the point patterns. Rotations may be better for this purpose than horizontal and vertical translations. The SADIE approach for examining the association between quadrat counts can also be used when the exact locations of points are known.

- A Mantel matrix randomization test can be used to test for spatial autocorrelation using data on a variable measured at a number of locations in a study area.

- A variogram can be used to summarise the nature of the spatial correlation for a variable that can be measured anywhere over a study area, using data from a sample of locations. A variogram cloud is a plot of $0.5(y_i - y_j)^2$ against the distance from point i to point j, for all pairs of sample points, where y_i is the value of the variable at the ith sample point. The experimental or empirical variogram is a curve that represents the average observed relationship between $0.5(y_i - y_j)^2$ and distance. A model variogram is a mathematical equation fitted to the experimental variogram.

- Kriging is a method for estimating the values of a variable at other than sampled points, using linear combinations of the values at known points. Before kriging is carried out a variogram model must be fitted using the available data.

- In the past Mantel matrix randomization tests have been used to examine whether two variables measured over space are associated, after allowing for any spatial correlation that might exist for the individual variables. There are potential problems with this application of the Mantel randomization test, which may be overcome by restricting randomizations to points located within spatially defined blocks.

- A geostatistical approach for testing for correlation between two variables X and Y involves comparing the observed correlation between the variables based on data from n sample locations with the distribution of correlations that is obtained from simulated data for which the two variables are independent, but each variable maintains the variogram that is estimated from the real data. Thus the generated correlations are for two variables that individually have spatial correlation structures like those for the real data, although they are actually uncorrelated.

CHAPTER 10

Censored Data

10.1 Introduction

Censored values occur in environmental data most commonly when the level of a chemical in a sample of material is less than the limit of quantitation (LOQ), or the limit of detection (LOD), where the meaning of LOQ and LOD depends on the methods being used to measure the chemical (Keith, 1991, Chapter 10). Censored values are generally reported as being less than detectable (LTD), with the detection limit (DL) specified.

There are questions raised by statisticians in particular about why censoring is done just because a measurement falls below the reporting limit, because an uncertain measurement is better than none at all (Lambert *et al.*, 1991). However, irrespective of these arguments it does seem that data values are inevitable in the foreseeable future in environmental data sets.

10.2 Single Sample Estimation

Suppose that there is a single random sample of observations, some of which are below the detection limit, DL. An obvious question then is how to estimate the mean and standard deviation of the population from which the sample was drawn. Some of the approaches that can be used are:

(a) With the simple substitution method the censored values are replaced by an assumed value. This might be zero, DL, DL/2, or a random value from a distribution over the range from zero to DL. After the censored values are replaced, the sample is treated as if it were complete to begin with. Obviously, replacing censored values by zero leads to a negative bias in estimating the mean, while replacing them with DL leads to a positive bias. Using random values from the uniform distribution over the range (0,DL) should give about the same estimated mean as is obtained from using DL/2, but gives a better estimate of the population variance (Gilliom and Helsel, 1986).

(b) Direct maximum likelihood methods are based on the original work of Cohen (1959). With these some distribution is assumed for the data and the likelihood function (which depends on both the observed and censored values) is maximized to estimate population parameters. Usually, a normal distribution is assumed, with the original data transformed to obtain this if necessary. These methods are well covered in the text by Cohen (1991).

(c) Regression on order statistics methods are alternatives to maximum likelihood methods that are easier to carry out in a spreadsheet, for example. One such approach works as follows for data from a normal distribution (Newman *et al.*, 1995). First, the n data values are ranked from smallest to largest, with those below the DL treated as the smallest. A normal probability plot is then constructed, with the ith largest data value (x_i) plotted against the normal score z_i, such that the probability of a value less than or equal to z_i is (i - 3/8)/(n + 1/4). Only the non-censored values can be plotted, but for these the plot should be approximately a straight line if the assumption of normality is correct. A line is fitted to the plot by ordinary linear regression methods. If this fitted line is x_i = a + bx_i, then the mean and standard deviation of the uncensored normal distribution are estimated by a and b, respectively. It may be necessary to transform the data to normality before this method is used, in which case the estimates a and b will need to be converted to the mean and standard deviation for untransformed data.

(d) With 'fill-in' methods, the complete data are used to estimate the mean and variance of the sampled distribution, which is assumed to be normal. The censored values are then set equal to their expected values based on the estimated mean and variance, and the resulting set of data treated as if it were a full set to begin with. The process can be iterated if necessary (Gleit, 1985).

(e) The robust parametric method is also a type of fill-in method. A probability plot is constructed, assuming either a normal or lognormal distribution for the data. If the assumed distribution is correct, then the uncensored observations should plot approximately on a straight line. This line is fitted by a linear regression, and extrapolated back to the censored observations, to give values for them. The censored values are then replaced by the values from the fitted regression line. If the detection limit varies, then this can be allowed for (Helsel and Cohn, 1988).

A computer program called UNCENSOR (Newman *et al.*, 1995) is available on the world wide web for carrying out eight different methods for estimating the censored data values in a sample, including versions of approaches (a) to (e) above. A program like this may be extremely useful as standard statistical packages seldom have these types of calculations as a standard menu option.

It would be convenient if one method for handling censored data was always best. Unfortunately, this is not the case. A number of studies have compared different methods, and it appears that in general for estimating the population mean and variance from a single random sample the robust parametric method is best when the underlying distribution of the data is uncertain, but if the distribution is known then maximum likelihood performs well, with an adjustment for bias with a sample size less than or equal to about 20 (Akritas *et al.*, 1994). In the manual for UNCENSOR, Newman *et al.* (1995) provide a flow chart for choosing a method that says more or less the same thing. On the other hand, in a manual on practical methods of data analysis the United States Environmental Protection Agency (1998) gives much simpler recommendations: with less than 15% of values censored replace these with DL, DL/2, or a small value; with between 15 and 50% of censored values use maximum likelihood, or estimate the mean excluding the same number of large values as small values; and with more than 50% of values censored, just base an analysis on the proportion of data values above a certain level.

See Akritas *et al.* (1994) for more information about methods for estimating means and standard deviations with multiple detection limits.

Example 10.1 A Censored Sample of 1,2,3,4-Tetrachlorobenzene

Consider the data shown in Table 10.1 for a sample of size 75 values of 1,2,3,4-tetrachlorobenzene (TcCB) in parts per million, from a possibly contaminated site. This sample has been used before in Example 1.7, and the original source was Gilbert and Simpson (1992, p. 6.22). For the present example it is modified by censoring any values less than 0.25, which are shown in Table 10.1 as '<0.25'. In fact, this means these values could be anywhere from 0.00 to 0.24 to two decimal places, so the detection limit is considered to be DL = 0.24.

Table 10.1 Measurements of TcCB (parts per thousand million) from a possibly contaminated site, with censoring of values less than 0.25

1.33	<0.25	<0.25	0.28	<0.25	<0.25	<0.25	0.47	<0.25	<0.25	<0.25	<0.25
18.40	<0.25	<0.25	<0.25	<0.25	<0.25	<0.25	168.6	<0.25	0.25	0.25	<0.25
0.48	0.26	5.56	<0.25	0.29	0.31	0.33	3.29	0.33	0.34	0.37	0.25
2.59	0.39	0.40	0.28	0.43	6.61	0.48	<0.25	0.49	0.51	0.51	0.38
0.92	0.60	0.61	0.43	0.75	0.82	0.85	<0.25	0.94	1.05	1.10	0.54
1.53	1.19	1.22	0.62	1.39	1.39	1.52	0.33	1.73	2.35	2.46	1.10
51.97	2.61	3.06									

For the uncensored data the sample mean and standard deviation are 4.02 and 20.27. It is interesting to see how well these values can be recovered from the censored data with some of the methods in general use.

First, consider the simple substitution methods. Replacing all of the censored values by zero, DL/2 = 0.12, DL = 0.24, and a uniform random value in the interval from 0.00 to 0.24 gave the following results for the sample mean and standard deviation (SD): replacement 0.00, mean = 3.97, SD = 20.28; replacement 0.12, mean = 4.00, SD = 20.28; replacement 0.24, mean = 4.03, SD = 20.27; and replacement uniform, mean = 4.00, SD = 20.28. Clearly in this example these simple substitution methods all work very well.

Newman *et al.*'s (1995) computer program UNCENSOR was used to calculate maximum likelihood estimates of the population mean and standard deviation using Cohen's (1959) method. The distribution was assumed to be lognormal because of the skewness indicated by three very large values. This gives the estimated mean and standard deviation to be 1.74 and 8.35, respectively. Using Schneider's (1986, Section 4.5) method for bias correction, the estimated mean and standard deviation change to 1.79 and 9.27, respectively. These maximum likelihood estimates are rather poor, in the sense that they differ very much from the estimates from the uncensored sample.

The regression on order statistics method can also be applied assuming a lognormal distribution, and it becomes apparent using this method that the assumption of a lognormal distribution is questionable. The calculations are shown in Table 10.2, and Figure 10.1 shows a normal probability plot for the logarithms of the uncensored values, i.e., the $\log_e(X)$ values against the normal scores Z. The data should plot approximately on a straight line if the logarithms of the TcCB concentrations are normally distributed. In fact, the plot appears to be curved, with the largest and smallest values being above the fitted straight line, showing that they are larger than expected for a normal distribution.

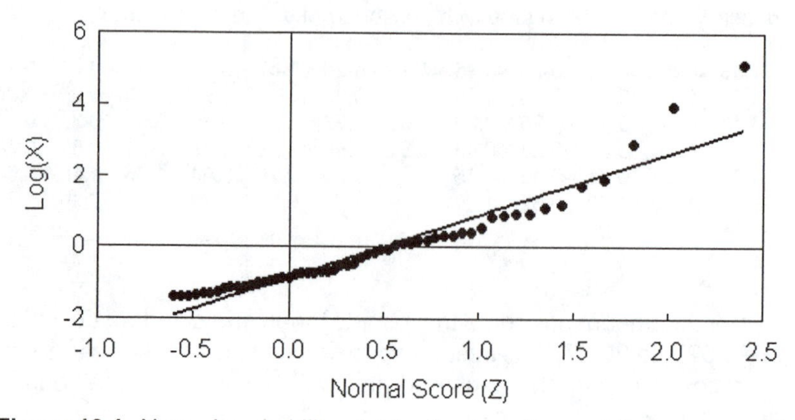

Figure 10.1 Normal probability plot for the logarithms of the uncensored TcCB concentrations, with a straight line fitted by ordinary regression methods.

Ignoring the possible problem with the assumed type of distribution, the equation of the fitted line shown in Figure 10.1 is $\log_e(X) = -0.83 + 1.75\,Z$. The estimated mean and standard deviation for the log-transformed data are therefore -0.83 and 1.75, respectively. To produce estimates of the corresponding mean and variance for the original distribution of TcCB concentrations is not now all that straightforward. As a quick approximation, equations (4.15) and (4.16) can be used. Thus the estimated mean is

$$E(X) = \exp(\mu + \tfrac{1}{2}\sigma^2) \approx \exp(-0.83 + 0.5\text{x}1.75^2) = 2.01$$

and the estimated variance is

$$Var(X) = \exp(2\mu + \sigma^2)\{\exp(\sigma^2) - 1\}$$

$$\approx \exp\{2\text{x}(-0.83) + 1.75^2\}\{\exp(1.75^2) - 1\} = 81.58,$$

so that the estimated standard deviation of TcCB concentrations is $\sqrt{81.58} = 9.03$.

Table 10.2 Calculations for the regression on order statistics with the censored TcCB data arranged in order from the smallest values (the censored ones) to the largest values.

Order (i)	P_i[1]	Z_i	X_i	$Log_e(X_i)$	Fitted[2]	Order (i)	P_i[1]	Z_i	X_i	$Log_e(X_i)$	Fitted[2]
1	0.01	-2.40	<0.25		-5.01	39	0.51	0.03	0.47	-0.76	-0.77
2	0.02	-2.02	<0.25		-4.36	40	0.53	0.07	0.48	-0.73	-0.71
3	0.03	-1.81	<0.25		-3.99	41	0.54	0.10	0.48	-0.73	-0.65
4	0.05	-1.66	<0.25		-3.73	42	0.55	0.13	0.49	-0.71	-0.59
5	0.06	-1.54	<0.25		-3.52	43	0.57	0.17	0.51	-0.67	-0.53
6	0.07	-1.44	<0.25		-3.34	44	0.58	0.20	0.51	-0.67	-0.48
7	0.09	-1.35	<0.25		-3.19	45	0.59	0.24	0.54	-0.62	-0.42
8	0.10	-1.27	<0.25		-3.05	46	0.61	0.27	0.60	-0.51	-0.36
9	0.11	-1.20	<0.25		-2.93	47	0.62	0.30	0.61	-0.49	-0.30
10	0.13	-1.14	<0.25		-2.81	48	0.63	0.34	0.62	-0.48	-0.23
11	0.14	-1.07	<0.25		-2.70	49	0.65	0.38	0.75	-0.29	-0.17
12	0.15	-1.02	<0.25		-2.60	50	0.66	0.41	0.82	-0.20	-0.11
13	0.17	-0.96	<0.25		-2.51	51	0.67	0.45	0.85	-0.16	-0.05
14	0.18	-0.91	<0.25		-2.42	52	0.69	0.48	0.92	-0.08	0.02
15	0.19	-0.86	<0.25		-2.33	53	0.70	0.52	0.94	-0.06	0.09
16	0.21	-0.81	<0.25		-2.25	54	0.71	0.56	1.05	0.05	0.15
17	0.22	-0.77	<0.25		-2.17	55	0.73	0.60	1.10	0.10	0.22
18	0.23	-0.73	<0.25		-2.09	56	0.74	0.64	1.10	0.10	0.29
19	0.25	-0.68	<0.25		-2.02	57	0.75	0.68	1.19	0.17	0.36
20	0.26	-0.64	<0.25		-1.95	58	0.77	0.73	1.22	0.20	0.44
21	0.27	-0.60	0.25	-1.39	-1.88	59	0.78	0.77	1.33	0.29	0.52
22	0.29	-0.56	0.25	-1.39	-1.81	60	0.79	0.81	1.39	0.33	0.60
23	0.30	-0.52	0.25	-1.39	-1.74	61	0.81	0.86	1.39	0.33	0.68
24	0.31	-0.48	0.26	-1.35	-1.67	62	0.82	0.91	1.52	0.42	0.76
25	0.33	-0.45	0.28	-1.27	-1.61	63	0.83	0.96	1.53	0.43	0.86
26	0.34	-0.41	0.28	-1.27	-1.54	64	0.85	1.02	1.73	0.55	0.95
27	0.35	-0.38	0.29	-1.24	-1.48	65	0.86	1.07	2.35	0.85	1.05
28	0.37	-0.34	0.31	-1.17	-1.42	66	0.87	1.14	2.46	0.90	1.16
29	0.38	-0.30	0.33	-1.11	-1.36	67	0.89	1.20	2.59	0.95	1.27
30	0.39	-0.27	0.33	-1.11	-1.30	68	0.90	1.27	2.61	0.96	1.40
31	0.41	-0.24	0.33	-1.11	-1.24	69	0.91	1.35	3.06	1.12	1.54
32	0.42	-0.20	0.34	-1.08	-1.18	70	0.93	1.44	3.29	1.19	1.69
33	0.43	-0.17	0.37	-0.99	-1.12	71	0.94	1.54	5.56	1.72	1.87
34	0.45	-0.13	0.38	-0.97	-1.06	72	0.95	1.66	6.61	1.89	2.08
35	0.46	-0.10	0.39	-0.94	-1.00	73	0.97	1.81	18.40	2.91	2.34
36	0.47	-0.07	0.40	-0.92	-0.94	74	0.98	2.02	51.97	3.95	2.70
37	0.49	-0.03	0.43	-0.84	-0.89	75	0.99	2.40	168.6	5.13	3.36
38	0.50	0.00	0.43	-0.84	-0.83						

[1]The $P_i = (i - 3/8)/(n + 1/4)$ are the probabilities used for calculating the Z scores, i.e. the probability of a value less than or equal to Z_i is P_i for the ith order statistic.
[2]The fitted values come from the fitted regression line shown in Figure 10.1. They are only used for the robust parametric method.

A better approach is to use the bias corrected method that is incorporated into UNCENSOR, which is based on a series expansion due to Finney (1941), and takes into account the sample size. For the example data, this gives the estimated mean and standard deviation of TcCB concentrations to be 1.92 and 15.66, respectively. Compared to the mean and standard deviation for the uncensored sample of 4.02 and 20.27, respectively, the regression on order statistics estimates without a bias correction are very poor, and not much better with a bias correction. Presumably this is because of the lack of fit of the lognormal distribution to the non-censored data (Figure 10.1).

Gleit's (1985) iterative fill-in method is another option in UNCENSOR. This gives the estimated mean and variance of TcCB concentrations to be 1.92 and 15.66, respectively. These are the same as the estimates obtained from the bias corrected regression on order statistics method, so are again rather poor.

Finally, consider the robust parametric method. This starts off the same way as the regression on order statistics method, with a probability plot of the data after a logarithmic transformation, with a fitted regression line (Figure 10.1). However, now instead of using the regression line to estimate the mean and variance of the fitted distribution, this line is extrapolated to obtain expected values for the censored data values, as shown in Figure 10.2. For example, the expected value for the smallest value in the sample is -5.0, corresponding to a normal score of -2.4, the second smallest value is -4.4, corresponding to a normal score of -2.0, and so on. The column headed 'Fitted' in Table 10.2 gives these expected values for the order statistics. The robust parametric method simply consists of replacing the smallest 20 censored values for $\log_e(X)$ with these expected values.

Having obtained values to 'fill-in' for the censored values of $\log_e(X)$, these are untransformed to obtain values for X itself. The sample mean and variance can then be calculated in the normal way. The completed sample is shown in Table 10.3. The mean and variance are 3.99 and 20.28, respectively, which are almost exactly the same as the values for the real data without censoring.

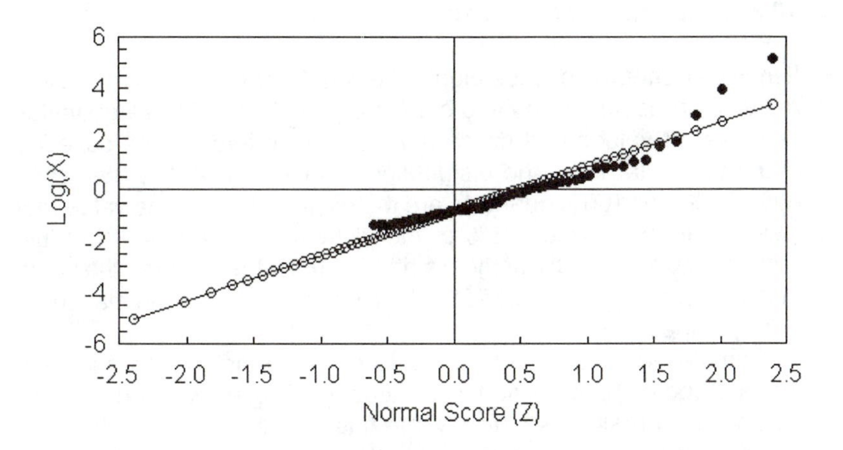

Figure 10.2 The regression line from Figure 10.1 extrapolated to estimate the censored values of the logarithm of TcCB values (● denotes an observed value of $\log_e(X)$, and ○ denotes an expected value from the regression line).

Too much should not be concluded from just one example. However, the simple substitution methods and the robust parametric method have very definitely worked better than the alternatives here for two reasons. First, the lognormal assumption is questionable for the methods that require this, other than the robust method. Second, the censored values are all very low and as long as they are replaced by any value below the detection limit the sample mean and standard deviation will be close to the values from the uncensored sample.

Table 10.3 The completed sample for the robust parametric method, with the filled-in values underlined

1.33	0.04	0.09	0.28	0.08	0.11	0.07	0.47	0.14	0.12	0.07	0.04
18.40	0.02	0.02	0.01	0.01	0.03	0.05	168.6	0.11	0.25	0.25	0.06
0.48	0.26	5.56	0.05	0.29	0.31	0.33	3.29	0.33	0.34	0.37	0.25
2.59	0.39	0.40	0.28	0.43	6.61	0.48	0.10	0.49	0.51	0.51	0.38
0.92	0.60	0.61	0.43	0.75	0.82	0.85	0.13	0.94	1.05	1.10	0.54
1.53	1.19	1.22	0.62	1.39	1.39	1.52	0.33	1.73	2.35	2.46	1.10
51.97	2.61	3.06									

10.3 Estimation of Quantiles

It may be better to describe highly skewed distributions with quantiles rather than using means and standard deviations. These quantiles are a set of values that divide the distribution into ranges covering equal percentages of the distribution. For example, the 0%, 25%, 50%, 75% and 100% quantiles are the minimum value, the value that just equals or exceeds 25% of the distribution, the value that just equals or exceeds 50% of the distribution (i.e., the median), the value that just equals or exceeds 75% of the distribution, and the maximum value, respectively.

Sample quantiles can be used to estimate distribution quantiles that are above the detection limit, although Akritas *et al.* (1994) note that simulation studies indicate that this can lead to bias when the quantiles are close to this limit. It is therefore better to use a parametric maximum likelihood approach when the distribution is known. When the distribution is uncertain, the robust parametric method can be used to 'fill-in' the censored data in the sample, before evaluating the sample quantiles as estimates of those for the underlying distribution of the data.

Distribution quantiles can be estimated with multiple detection limits. See Akritas *et al.* (1994, Section 2.6) for more details.

10.4 Comparing the Means of Two or More Samples

The comparison of the means of two or more samples is complicated with censored data, particularly if there is more than one detection limit. The simplest approach involves just replacing censored data by zero, DL, or DL/2, and then using standard methods either to test for a significant mean difference or to produce a confidence interval for the mean difference between the two sampled populations. In fact, this approach seems to work quite well, and based on a simulation study of ten alternative ways for handling censoring suggests that a good general strategy involves substituting DL for censored values when up to 40% of observations are censored, and substituting DL/2 when more than 40% of observations are censored (Clarke, 1994). However, this strategy is not always the best and the United States Environmental Protection Agency and United States Army Corps of Engineers (1998, Table D-12) give some more complicated rules that depend on the type of data, whether samples have equal variances, the coefficient of variation, and the type of data distribution.

When it can be assumed that the data come from a particular distribution, comparisons between groups can be based on the method of maximum likelihood, as described by Dixon (1998). One of the advantages of maximum likelihood estimation is the approximate variances and covariances of the estimators that are available. Using these it is possible to carry out a large sample test for whether the estimated population means are significantly different, or to find an approximate confidence interval for this difference.

For small samples, Dixon (1998) suggests the use of bootstrap methods for hypothesis testing and producing confidence intervals, as discussed further in the following example. This has obvious generalizations for use with other data distributions, and with more than two samples. Dixon also discusses the use of non-parametric methods for comparing samples, and the use of equivalence tests with data containing censored values.

Example 10.2 Upstream and Downstream Samples

The data from one of the examples considered by Dixon (1998) are shown in Table 10.4. The variable being considered is the dissolved orthophosphate concentration (DOP, mg/l) measured for water from the Savannah River in South Carolina, USA. One sample is of 41 observations taken upstream of a potential contamination source, and the second sample is of 42 observations taken downstream. A higher general level of DOP downstream is clearly an indication that contamination has occurred. There are three DL values in this example, <1, <5, and <10, which occurred because the DL depends on dilution factors and other aspects of the chemical analysis that changed during the study.

The number of censored observations is high, consisting of 26 in each of the samples, and 63% of the values overall. Given the high detection limit of 10 for some of the data, simple substitution methods seem definitely questionable here, and an analysis assuming a parametric distribution seems like the only reasonable approach.

Table 10.4 Dissolved orthophosphate concentrations in samples upstream and downstream of a possible source of contamination, with three different detection limits

Sample 1, Upstream of Possible Contamination Source											
1	2	4	3	3	<10	2	<10	<5	<10	<5	3
<5	<5	<10	<5	<10	<1	<10	7	<5	<1	<5	2
<10	5	5	<5	<10	<1	<5	<10	<5	14	5	2
<10	<10	7	<1	<10							

Sample 2, Downstream of Possible Contamination Source											
4	<5	<1	4	3	9	<10	4	<5	<10	<10	8
<10	3	<5	<5	<10	5	<5	<10	6	<5	1	4
<10	<5	<5	<10	5	4	2	<5	<10	<5	<10	<5
<1	<10	4	<5	20	<10						

Dixon (1998) assumed that the data values X are lognormally distributed, with $\log_e(X)$ having the same variance upstream and downstream of the potential source of contamination. On this basis he obtained the following maximum likelihood estimates: mean DOP upstream, 0.73 with standard error 0.19; mean DOP downstream, 1.02 with standard error 0.17; mean difference between downstream and upstream, 0.24 with standard error 0.23. This clearly indicates that the two samples could very well come from the same lognormal distribution.

Dixon also applied parametric bootstrap methods for testing for a significant mean difference between the upstream and downstream samples, and for finding confidence intervals for the mean difference between downstream and upstream. The adjective 'parametric' is used here because samples are taken from a specific parametric distribution (the lognormal) rather than just resampling the data with replacement as explained in Section 4.7. These bootstrap methods are more complicated than the usual maximum likelihood approach but do have the advantage of being expected to have better properties with small sample sizes.

The general approach proposed for hypothesis testing with two samples of size n_1 and n_2 is:

(a) Estimate the overall mean and standard deviation assuming no difference between the two samples. This is the null hypothesis distribution.

(b) Draw two random samples with sizes n_1 and n_2 from a lognormal distribution with the estimated mean and standard deviation, censoring these using the same detection limits as applied with the real data.

(c) Use maximum likelihood to estimate the population means μ_1 and μ_2 by $\hat{\mu}_1$ and $\hat{\mu}_2$, and to approximate the standard error $SE(\hat{\mu}_2 - \hat{\mu}_1)$ of the difference.

(d) Calculate the test statistic

$$T = (\hat{\mu}_2 - \hat{\mu}_1)/SE(\hat{\mu}_2 - \hat{\mu}_1),$$

where $SE(\hat{\mu}_2 - \hat{\mu}_1)$ is the estimated standard error.

(e) Repeat steps (b) to (d) many times to generate the distribution of T when the null hypothesis is true, and declare the observed value of T for the real data to be significantly large at the 5% level if it exceeds 95% of the computer generated values.

Other levels of significance can be used in the obvious way. For example, significance at the 1% level requires the value of T for the real data to exceed 99% of the computer generated values. For a two-sided test the test statistic T just needs to be changed to

$$T = |\hat{\mu}_2 - \hat{\mu}_1|/SE(\hat{\mu}_2 - \hat{\mu}_1),$$

so that large values of T occur with either large positive or large negative differences between the sample means.

For the DOP data the observed value of T is 0.24/0.23 = 1.04. As could have been predicted, this is not at all significantly large with the bootstrap test, for which it was found that 95% of the computer-generated T values were less than 1.74.

The bootstrap procedure for finding confidence intervals for the mean difference uses a slightly different algorithm. See Dixon's (1998) paper for more details. The 95% confidence interval for the DOP mean difference was found to be from -0.24 to +0.71.

10.5 Regression with Censored Data

There are times when it is desirable to fit a regression equation to data with censoring. For example, in a simple case it might be assumed that the usual simple linear regression model

$$Y_i = \alpha + \beta X_i + \epsilon_i$$

holds, but either some of the Y values are censored, or both X and Y values are censored.

There are a number of methods available for estimating the regression parameters in this type of situation, including maximum likelihood approaches that assume particular distributions for the error term, and a range of non-parametric methods that avoid making such assumptions. For more information, see the reviews by Schneider (1986, Chapter 5) and Akritas *et al.* (1994).

10.6 Chapter Summary

- Censored values most commonly occur in environmental data when the level of a chemical in a sample of material is less than what can be reliably measured by the analytical procedure. Censored values are generally reported as being less than the detection limit (DL).

- Methods for handling censored data for the estimation of the mean and standard deviation from a single sample include (a) the simple substitution of zero, DL, DL/2 or a random value between zero and DL for censored values to complete the sample; (b) maximum likelihood methods, assuming that data follow a specified parametric distribution; (c) regression on order statistics methods, where the mean and standard deviation are estimated by fitting a linear regression line to a probability plot; (d) fill-in methods, where the mean and standard deviation are estimated from the uncensored data and then used to predict the censored values to complete the sample; and (e) robust parametric methods, which are similar to the regression on order statistic methods except that the fitted regression line is used to predict the censored values in order to complete the sample.

- No method for estimating the mean and standard deviation of a single sample is always best. However, the robust parametric method is often best if the underlying distribution of data is uncertain, and maximum likelihood methods (with a bias correction for small samples) are likely to be better if the distribution is known.

- An example shows good performance of the simple substitution methods and a robust parametric method, but poor performance of other methods when a distribution is assumed to be lognormal, when this is apparently not true.

- It may be better to describe highly skewed distributions by sample quantiles (values that exceed defined percentages of the distribution) rather than means and standard deviations. Estimation of the quantiles from censored data is briefly discussed.

- For comparing the means of two or more samples subject to censoring it may be reasonable to use simple substitution to complete samples. Alternatively, maximum likelihood can be used, possibly assuming a lognormal distribution for data.

- An example involving the comparison of two samples from upstream and downstream of a potential source of contamination is described. Maximum likelihood is used to estimate population parameters of assumed lognormal distributions, with bootstrap methods used to test for a significant mean difference, and to produce a confidence interval for the true mean difference.

- Regression analysis with censored data is briefly discussed.

CHAPTER 11

Monte Carlo Risk Assessment

11.1 Introduction

Monte Carlo simulation for risk assessment is a relatively new idea, made possible by the increased computer power that has become available to environmental scientists in recent years. The essential idea is to take a situation where there is a risk associated with a certain variable, such as an increased incidence of cancer when there are high levels of a chemical in the environment. The level of the chemical is then modelled as a function of other variables, some of which are random variables, and the distribution of the variable of interest is generated through a computer simulation. It is then possible, for example, to determine the probability of the variable of interest exceeding an unacceptable level. The description 'Monte Carlo' comes from the analogy between a computer simulation and repeated gambling in a casino.

The basic approach for Monte Carlo methods involves five steps:

- A model is set up to describe the situation of interest.
- Probability distributions are assumed for input variables, such as chemical concentrations in the environment, ingestion rates, exposure frequency, etc.
- Output variables of interest are defined, such as the amounts of exposure from different sources, the total exposure, etc.).
- Random values from the input distributions are generated for the input variables, and the resulting output distributions are derived.
- The output distributions are summarised by statistics such as the mean, the value exceeded 5% of the time, etc.

There are three main reasons for using Monte Carlo methods:

(1) The alternative is often to assume the worse possible case for each of the input variables contributing to an output variable of

interest. This can then lead to absurd results, such as the Record of Decision for a US Superfund site at Oroville, California, which specifies a clean-up goal of 5.3×10^{-7} µg/litre for dioxin in groundwater, which is about 100 times lower than the drinking water standard and 20 times lower than current limits of detection (United States Environmental Protection Agency, 1989b). Thus there may be unreasonable estimates of risk, and unreasonable demands for action associated with those risks, leading to the questioning of the whole process of risk assessment.

(2) Properly conducted, a probabilistic assessment of risk gives more information than a deterministic assessment. For example, there may generally be quite low exposure to a toxic chemical, but occasionally individuals may get extreme levels. It is important to know this, and in any case the world is stochastic rather than deterministic, so deterministic assessments are inherently unsatisfactory.

(3) Given that a probability-based assessment is to be carried out, the Monte Carlo approach is usually the easiest way to do this.

On the other hand, Monte Carlo methods are only really needed when the 'worse case' deterministic scenario suggests that there may be a problem. This is because making a scientifically defensible Monte Carlo analysis, properly justifying assumptions, is liable to take a great deal of time.

11.2 Principles for Monte Carlo Risk Assessment

The United States Environmental Protection Agency has put considerable effort into the development of reasonable approaches for using Monte Carlo simulation. Their website on this topic (United States Environmental Protection Agency, 1999) is full of useful information, as is their policy document (United States Environmental Protection Agency, 1997) that can be obtained from the same source.

In the policy document the following guiding principles are stated for Monte Carlo studies:

- The purpose and scope should be clearly explained in a 'problem formulation'.

- The methods used (models, data, assumptions) should be documented and easily located with sufficient detail for all results to be reproduced.

- Sensitivity analyses should be presented and discussed.

- Correlations between input variables should be discussed and accounted for.

- Tabular and graphical representation of input and output distributions should be provided.

- The means and upper tails of output distributions should be presented and discussed.

- Deterministic and probabilistic estimates should be presented and discussed.

- The results from output distributions should be related to reference doses, reference concentrations, etc.

It is stressed that this is a minimum set of principles that are not intended to restrict the use of new scientifically defensible methods.

11.3 Risk Analysis Using a Spreadsheet Add-On

For many applications, the simplest way to carry out a Monte Carlo risk analysis is using a spreadsheet add-on. Two such add-ons are @Risk (Palisade Corporation, 2000), and Crystal Ball (Decisioneering Inc., 2000). In both cases these products use the spreadsheet as a basis for calculations, adding extra facilities for simulation. Typically, what is done is to set up the spreadsheet with one or more random input variables and one or more output variables that are functions of the input variables. Each recalculation of the spreadsheet yields new random values for the input variables, and consequently new random values for the output variables. What @Risk and Crystal Ball do is to allow the recalculation of the spreadsheet hundreds or thousands of times, followed by the automatic generation of tables and graphs that summarise the characteristics of the output distributions. The following example illustrates the general procedure.

Example 11.1 Contaminant Uptake Via Tapwater Ingestion

This example concerns cancer risks associated with tapwater ingestion of Maximum Contaminant Levels (MCL) of tetrachloroethylene in high risk living areas. It is a simplified version of a case study considered by Finley *et al.* (1993).

A crucial equation gives the dose of tetrachloroethylene received by an individual (mg/kg-day) as a function of other variables. This equation is

$$\text{Dose} = (C \times IR \times EF \times ED)/(BW \times AT) \qquad (11.1)$$

where C is the chemical concentration in the tapwater (mg/litre), IR is the ingestion rate of water (litres/day), EF is the exposure frequency (days/year), ED is the exposure duration (years), BW is the body weight (kg), and AT is the averaging time (days). The numerator is the total dose received in EF x ED exposure days, while the denominator is the total number of days in the period considered. Dose is therefore the average daily dose of tetrachloroethylene per kilogram of body weight. The aim in this example is to determine the distribution of this variable over the population of adults living in a high risk area.

The variables on the right-hand side of equation (11.1) are the input variables for the study. These are assumed to have the following characteristics:

C, the chemical concentration, is assumed to be constant at the MCL for the chemical of 5 µg/litre;

IR, the ingestion rate of tapwater, is assumed to have a mean of 1.1 and a range of 0.5-5.5 litres per day, based on survey data;

EF, the exposure frequency, is set at the United States Environmental Protection Agency upper point estimate of 350 days per year;

ED, the exposure duration, is set at 12.9 years based on the average residency tenure in a household in the United States;

BW, the body weight is assumed to have a uniform distribution between 46.8 (5th percentile female in the United States) and 101.7 kg (95th percentile male in the United States); and

AT, the averaging time, is set at 25,550 days (70 years).

Thus C, EF, ED and AT are taken to be constants, while IR and BW are random variables. It is, of course, always possible to argue with the assumptions made with a model like this. Here it suffices to say that the constants appear to be reasonable values, while the distributions for the random variables were based on survey results. For IR the empirical distribution shown in Table 11.1 is used because this gives the correct mean and range.

Table 11.1 Distribution used for the ingestion rate of tapwater for the individuals living in high risk areas

Ingestion rate (l/day)	Probability
0.50	0.2857
0.75	0.2571
1.00	0.2286
1.50	0.0857
2.00	0.0571
2.50	0.0286
3.00	0.0143
3.50	0.0143
4.00	0.0114
4.50	0.0086
5.00	0.0057
5.50	0.0029
Total	1.0000

There are two output variables:

Dose, the dose received (mg/kg-day) as defined before; and

ICR, the increased cancer risk (the increase in the probability of a person getting cancer) which is set at Dose x CPF(oral), where CPF(oral) is the cancer potency factor for the chemical taken orally.

For the purpose of the example CPF(oral) was set at the United States Environmental Protection Agency's upper limit of 0.051.

A spreadsheet was set up containing dose and ICR as functions of the other variables, with the @Risk add-on activated. Each recalculation of the spreadsheet then produced new random values for IR and BW, and consequently for dose and ICR, to simulate the situation for a random individual from the population at risk. The number of simulated sets of data was set at 10,000. Table 11.2

shows some of the summary output obtained (minimums, maximums, means, etc.), while Figure 11.1 shows the distribution obtained for the ICR. (The dose distribution is the same but with the horizontal axis divided by 0.051.)

The 50th and 95th percentiles for the ICR distribution are 0.05×10^{-5} and 0.20×10^{-5}, respectively. Finlay *et al.* (1993) note that the 'worse case' scenario gives an ICR of 0.53×10^{-5}, but a value this high was never seen with the 10,000 simulated random individuals from the population at risk. Hence the 'worse case' scenario actually represents an extremely unlikely event. At least, this is the case based on the assumed model.

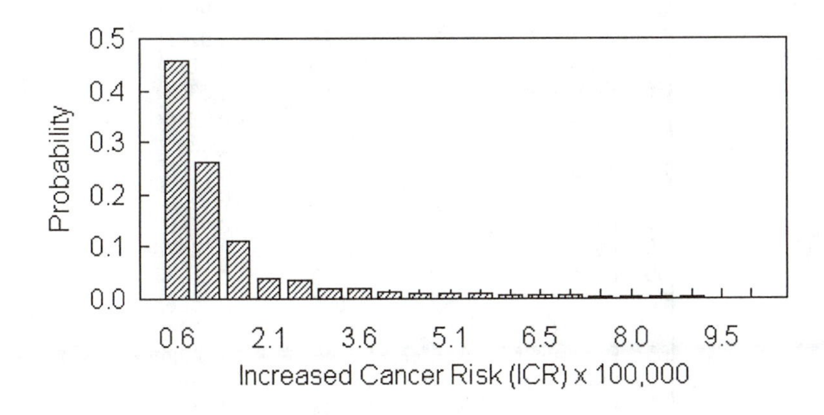

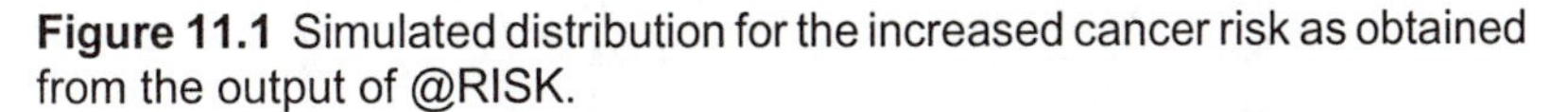

Figure 11.1 Simulated distribution for the increased cancer risk as obtained from the output of @RISK.

Table 11.2 Summary of output from @Risk based on simulating 10,000 random individuals from the population living in a high contamination area

Name	Dose*10^{-5}	ICR*10^{-5}
Description	Output	Output
Cell	A35	C35
Minimum =	0.4344	0.0222
Maximum =	10.2119	0.5208
Mean =	1.3778	0.0703
Std Deviation =	1.1539	0.0588
Variance =	1.3314	0.0035
Skewness =	2.7076	2.7076
Kurtosis =	11.9364	11.9364
Errors Calculated =	0	0
Mode =	1.1234	0.0573
5% Perc =	0.4795	0.0245
10% Perc =	0.5348	0.0273
15% Perc =	0.6042	0.0308
20% Perc =	0.6671	0.0340
25% Perc =	0.7069	0.0361
30% Perc =	0.7560	0.0386
35% Perc =	0.8141	0.0415
40% Perc =	0.8748	0.0446
45% Perc =	0.9161	0.0467
50% Perc =	0.9746	0.0497
55% Perc =	1.0509	0.0536
60% Perc =	1.1452	0.0584
65% Perc =	1.2516	0.0638
70% Perc =	1.3708	0.0699
75% Perc =	1.5357	0.0783
80% Perc =	1.7690	0.0902
85% Perc =	2.1374	0.1090
90% Perc =	2.6868	0.1370
95% Perc =	3.8249	0.1951

11.4 Further Information

A good starting point for more information is the Risk Assessment Forum home page (United States Environmental Protection Agency, 2000). For examples of a range of applications of Monte Carlo methods, a special 400 page issue of the journal *Human and Ecological Risk Assessment* will be useful (Association for the Environmental Health of Soils, 2000). For more information about @Risk, see the book by Winston (1996).

11.5 Chapter Summary

- The Monte Carlo method uses a model to generate distributions for output variables from assumed distributions for input variables.

- These methods are useful because 'worse case' deterministic scenarios may have a very low probability of ever occurring, stochastic models are usually more realistic, and Monte Carlo is the easiest way to use stochastic models.

- The guiding principles of the United States Environmental Protection Agency for Monte Carlo analysis are summarised.

- An example is provided to show how Monte Carlo simulation can be done with the @RISK add-on for spreadsheets.

- Sources of further information are noted.

CHAPTER 12

Final Remarks

There are a number of books available describing interesting applications of statistics in environmental science. The book series *Statistics in the Environment* is a good starting point because it contains papers arising from conferences with different themes covering environmental monitoring, pollution and contamination, climate change and meteorology, water resources, fisheries and forestry, radiation, and air quality (Barnett and Turkman, 1993, 1994, 1997). Further examples of applications are also provided by Fletcher and Manly (1994), Fletcher *et al.* (1998), and Nychka *et al.* (1998).

For more details about statistical methods in general, the handbook edited by Patil and Rao (1994) or the *Encyclopedia of Environmetrics* (El-Shaarawi and Piegorsch, 2001) are good general references.

There are several journals that specialize in publishing papers on applications of statistics in environmental science, with the most important being *Environmetrics*, *Ecological and Environmental Statistics*, and *The Journal of Agricultural, Biological and Environmental Statistics*. In addition, journals on environmental management frequently contain papers on statistical methods.

It is always risky to attempt to forecast the development of a subject area. No doubt new statistical methods will continue to be proposed in all of the areas discussed in this book, but it does seem that the design and analysis of monitoring schemes, time series analysis, and spatial data analysis will receive particular attention as far as research is concerned. In particular, approaches for handling temporal and spatial variation at the same time are still in the early stages of development.

One important topic that has not been discussed in this book is the handling of the massive multivariate data sets that can be produced by automated recording devices. Often the question is how to reduce the data set to a smaller (but still very large) set that can be analysed by standard statistical methods. There are many future challenges for the statistics profession in learning how to handle the problems involved (Manly, 2000).

Appendix A Some Basic Statistical Methods

A1 Introduction

It is assumed that readers of this book already have some knowledge of elementary statistical methods. This appendix is therefore not intended to be a full introduction to these methods. Instead, it is intended to be a quick refresher course for those who may have forgotten some of this material. Nevertheless, this appendix covers the minimum background needed for reading the rest of the book, so that those who have not formally studied statistics before may find it a sufficient starting point.

A2 Distributions for Sample Data

Random variation is the raw material of statistics. When observations are taken on an environmental variable they usually display variation to a greater or lesser extent. For example, Table A1 shows the values for 1,2,3,4-tetrachlorobenzene (TcCB) in parts per thousand million for 47 samples from an uncontaminated site used as a reference for comparison with a possibly contaminated site (Gilbert and Simpson, 1992, p. 6-22). These values vary from 0.22 to 1.33, presumably due to natural variation in different parts of the site, plus some analytical error involved in measuring a sample. As a subject, the main concern of statistics is to quantify this type of variation.

Table A1 Measurements of TcCB in parts per thousand million for 47 samples taken from different locations at an uncontaminated site

0.60	0.50	0.39	0.84	0.46	0.39	0.62	0.67	0.69	0.81	0.38	0.79
0.43	0.57	0.74	0.27	0.51	0.35	0.28	0.45	0.42	1.14	0.23	0.72
0.63	0.50	0.29	0.82	0.54	1.13	0.56	1.33	0.56	1.11	0.57	0.89
0.28	1.20	0.76	0.26	0.34	0.52	0.42	0.22	0.33	1.14	0.48	

Data distributions come in two basic varieties. When the values that can be observed are anything within some range then the distribution is said to be continuous. Hence the data shown in Table

A1 are continuous because in principle any value could have been observed over a range that extends down to 0.22 or less, and up to 1.33 or more. On the other hand, if only certain particular values can be observed, then the distribution is said to be discrete. An example is the number of lesions observed on the lungs of rats at the end of an experiment where they were exposed to a certain toxic substance for a certain period of time. In that case an individual rat can have a number of lesions equal to 0 or 1 or 2, and so on. Often the possible values for a discrete variable are counts like this.

There are many standard distributions for continuous data. Here only the normal distribution (also sometimes called the Gaussian distribution) is considered. This distribution is characterised as being 'bell-shaped', with most values being near the centre of the distribution. There are two parameters to describe the distribution, the mean and standard deviation, which are often denoted by μ and σ, respectively. There is also a function to describe the distribution in terms of these two parameters, which is referred to as a probability density function (pdf).

An example pdf is shown in Figure A1 for the distribution with μ = 5 and σ = 1. A random value from this normal distribution is one for which the probability of obtaining a particular value x is proportional to the height of the function. Thus 5 is the value that is most likely to occur and values outside the range 2 to 8 will occur rather rarely.

In general it turns out that for all normal distributions about 67% of values will be in the range $\mu \pm \sigma$, about 95% will be in the range $\mu \pm 2\sigma$, and about 99.7% in the range $\mu \pm 3\sigma$.

The normal distribution with $\mu = 0$ and $\sigma = 1$ is of special importance. This is called the standard normal distribution, and variables with this distribution are often called Z-scores. A table of probabilities for this particular distribution is in Table B1 in Appendix B because this is useful for various statistical analyses.

If a large sample of random values are selected independently from a normal distribution to give values $x_1, x_2, ..., x_n$, then the mean of these values

$$\bar{x} = (x_1 + x_2 + ... + x_n)/n \qquad (A1)$$

will be approximately equal to μ. Thus $\bar{x}$ is an estimate of μ. This is in fact true for all other distributions as well: the sample mean is said to be an 'estimator' of the distribution mean for a random sample from any distributions.

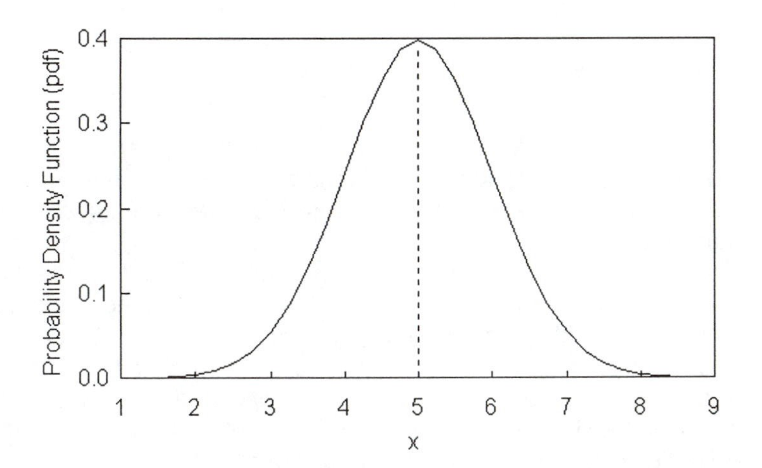

Figure A1 The probability density function (pdf) for the normal distribution with a mean of 5 and a standard deviation of 1.

The square of the standard deviation of the distribution, σ^2, is called the variance of the distribution. For the large sample this should be approximately equal to the sample variance, which is defined to be

$$s^2 = \{(x_1 - \bar{x})^2 + (x_2 - \bar{x})^2 + ... + (x_n - \bar{x})^2\}/\{n - 1\}, \quad \text{(A2)}$$

where the division by n - 1 (rather than n) is made to remove a tendency to under-estimate the population variance which would otherwise occur.

The square root of the sample variance, s, is called the sample standard deviation, and this should be close to σ for large samples. Hence, s^2 estimates σ^2, and s estimates σ. More generally, s^2 is an estimator of the distribution variance, and s is an estimator of the distribution standard deviation when calculated from a random sample from any distribution.

The sample mean and variance are often referred to using the more concise notation

$$\bar{x} = \sum_{i=1}^{n} x_i / n,$$

and

$$s^2 = \sum_{i=1}^{n} (x_i - \bar{x})^2 /(n - 1).$$

Here $\sum$ is the summation operator, indicating that elements following this sign are to be added up over the range from i = 1 to n. Consequently, the last two equations are exactly equivalent to equations (A1) and (A2), respectively.

The name 'normal' implies that the normal distribution is what will usually be found for continuous data. This is, however, not the case. It is a matter of fact that many naturally occurring distributions appear to be approximately normal, but certainly not all of them. Environmental variables often have a distribution that is skewed to the right rather than being bell-shaped. For example, a histogram for the TcCB values in Table A1 is shown in Figure A2. Here there is a suggestion that if many more data values were obtained from the site then the distribution would not quite be symmetrical about the mean, which is about 0.6. Instead, the right tail would extend further from the mean than the left tail.

There are also many standard distributions for discrete data. Here only the binomial distribution will be considered, where this owes its importance to its connection with data in the form of proportions. This distribution arises when there is a certain constant probability that an observation will have a certain property of interest. For example, a series of samples might be taken from random locations at a certain study site and there is interest in whether the level of a toxic chemical is higher than a specified value. If the probability of an exceedence is p for a randomly chosen location, then the probability of observing exactly x exceedences from a sample of n locations is given by the binomial distribution

$$P(x) = {}^nC_x p^x (1 - p)^{n-x}, \qquad \text{(A3)}$$

where the possible values of x are 0, 1, 2, ..., n. In this equation ${}^nC_x = n!/\{x!(n - x)!\}$ is the number of combinations of n things taken r at a time, with $a! = a(a - 1)(a - 2)...(2)(1)$ being the factorial function.

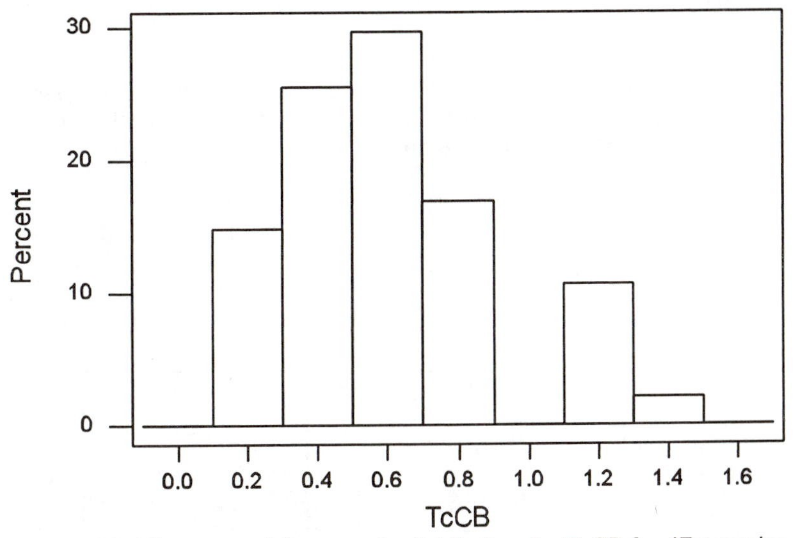

Figure A2 Histogram of the sample distribution for TcCB for 47 samples from a site, with the height of the bars proportional to the percentage of data values within the ranges that they cover.

The mean and variance of the binomial distribution are $\mu = np$ and $\sigma^2 = np(1 - p)$, respectively. If a large sample of values $x_1, x_2, \ldots x_n$ is taken from the distribution, then the sample mean $\bar{x}$ and the sample variance s^2 calculated using equations (A1) and (A2) will be approximately equal to μ and σ^2, respectively.

An example of a binomial distribution comes from the accidental bycatch of New Zealand sea lions *Phocarctos hookeri* during trawl fishing for squid around the Auckland Islands to the south of New Zealand. Experience shows that for any individual trawl the probability of catching a sea lion in the net is fairly constant at about 0.025, and there are about 3500 trawls in a summer fishing season (Manly and Walshe, 1999). It is extremely rare for more than one animal to be captured at a time. Suppose that the trawls during a particular season are considered in groups of 100, and the number of sea lions caught is recorded for trawls 1-100, 101-200, and so on up to 3401-3500. Then there would be 35 observations, each of which is a random value from a binomial distribution with n = 100 and p = 0.025. Table A2 shows the type of data that can be expected to be obtained under these conditions with a histogram in Figure A3.

The results shown in Table A2 come from a computer simulation of the sea lion bycatch. Such simulations are often valuable for obtaining an idea of the variation to be expected according to an assumed model for data.

Table A2 Simulated bycatch of New Zealand sea lions in 35 successive sets of 100 trawls each, where each observation is a random value from a binomial distribution with n = 100 and p = 0.025

1	1	2	1	3	2	1	3	2	2	2	2
3	0	2	0	3	2	1	0	1	4	2	3
3	2	4	1	4	2	2	2	1	2	3	

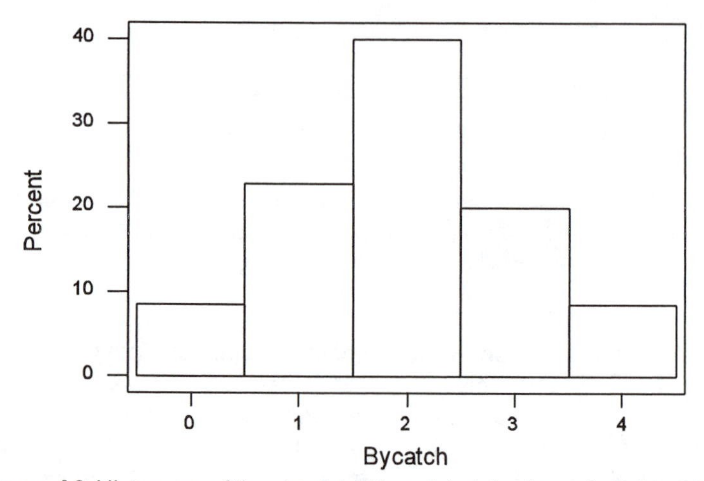

Figure A3 Histogram of the simulated bycatch data shown in Table A2, with the height of each bar reflecting the percentage of observations for the value concerned.

One of the reasons for the importance of the normal distribution is the way that some other distributions become approximately normal, given the right circumstances. This applies for the binomial distribution providing that the sample size n is large enough, and a good general rule is that a binomial distribution is similar to a normal distribution providing that the condition $np(1 - p) \geq 5$ applies. This result is particularly useful because of the fact that if a sample count follows a binomial distribution with mean np and variance $np(1 - p)$ then, providing that $np(1 - p) \geq 5$, the sample proportion x/n can reasonably be treated as coming from a normal distribution with mean p and variance $p(1 - p)/n$. This is a key result for the analysis of data consisting of observed proportions.

For the bycatch data $np(1 - p) = 100(0.025)(0.975) = 2.44$, which is much less than 5. Therefore the condition for a good normal approximation does not apply. Nevertheless, the distribution observed is quite symmetric and bell-shaped.

A3 Distributions of Sample Statistics

There are some distributions that arise indirectly as a result of calculating values that summarise samples, like the sample mean and the sample variance. These values that summarise samples are called sample statistics. Hence it is the distributions of these sample statistics that are of interest. The reason why it is necessary to know about these particular distributions is that they are needed for drawing conclusions from data, using tests of significance and confidence limits, for example.

The first distribution to consider for sample statistics is the t-distribution. This arises when a random sample of size n is taken from a normal distribution with mean μ. If the sample mean $\bar{x}$ and the sample variance s^2 are calculated using equations (A1) and (A2), then the quantity

$$t = (\bar{x} - \mu)/(s/\sqrt{n}) \quad \text{(A4)}$$

follows what is called a t-distribution with n - 1 degrees of freedom (df). For large values of n this distribution approaches a standard normal distribution with mean 0 and standard deviation 1.

It is important to realize what exactly it means to say that the quantity $(\bar{x} - \mu)/(s/\sqrt{n})$ follows a t-distribution. It does in the sense that if the process of taking a random sample of size n from the normal distribution was itself repeated a large number of times, then the resulting distribution obtained for the values of $(\bar{x} - \mu)/(s/\sqrt{n})$ would be the t-distribution. One sample of size n therefore yields a single random value from this distribution.

For small values of the df the t-distribution is more spread out than the normal distribution with mean 0 and standard deviation 1. This is illustrated in Figure A4, which shows the shape of the distribution for several values of the df. Table B2 in Appendix B gives what are called 'critical values' for t-distributions, because these will be useful later in this book. The values in this table are such that they are exceeded with a probability of 0.05, 0.025, 0.01 or 0.005. Note that because the distribution is symmetric about zero, $\text{Prob}(t < -c) = \text{Prob}(t > c)$, for any critical value c.

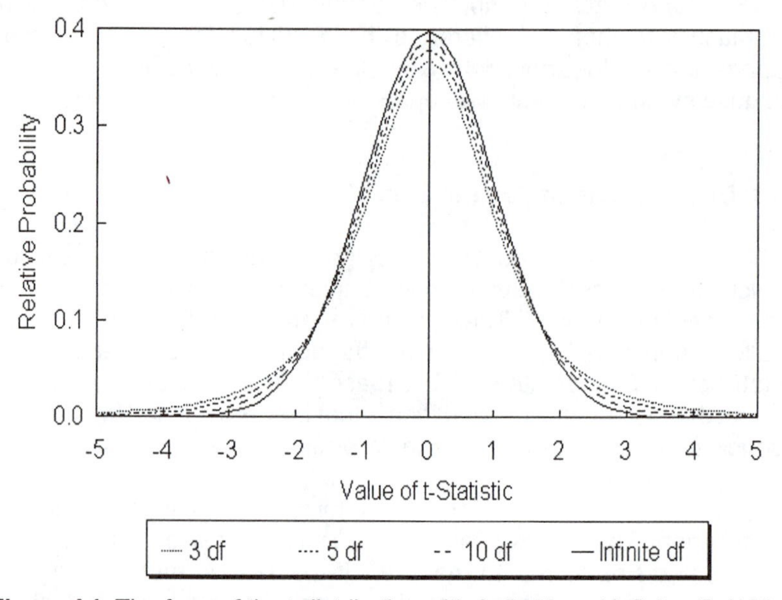

Figure A4 The form of the t-distribution with 3, 5, 10 and infinite df. With infinite df the t-distribution is the standard normal distribution, with a mean of 0 and a standard deviation of 1.

The second distribution to be considered is the chi-squared distribution. If a random sample of size n is taken from a normal distribution with variance σ^2, and the sample variance s^2 is calculated using equation (1.2), then the quantity

$$X^2 = (n - 1)s^2/\sigma^2 \quad \text{(A5)}$$

is a random value from the chi-squared distribution with n - 1 df. The shape of this distribution depends very much on the df, becoming more like a normal distribution as the df increases. Some example shapes are shown in Figure A5, and Table B3 in Appendix B gives critical values for the distribution that are needed for various purposes.

The third and last distribution to be considered is the F-distribution. Suppose that a random sample of size n_1 is taken from a normal distribution with variance σ^2, and the sample variance s_1^2 is calculated using equation (A2), and then a random sample of size n_2 is independently taken from a second normal distribution with the same variance, and the sample variance s_2^2 is calculated, again using equation (A2). In that case, the ratio of the sample variances,

$$F = s_1^2/s_2^2, \quad \text{(A6)}$$

will be a random value from the F-distribution with df $n_1 - 1$ and $n_2 - 1$. Like the chi-squared distribution, the F-distribution may take a variety of different shapes, depending on the df. Some examples are shown in Figure A6, and Table B4 in Appendix B gives critical values for the distribution that are needed later.

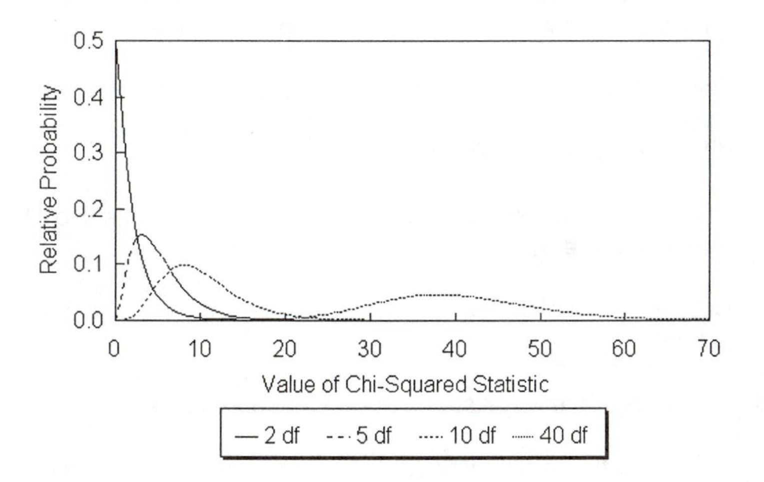

Figure A5 Form of the chi-squared distribution with 2, 5, 10 and 40 df.

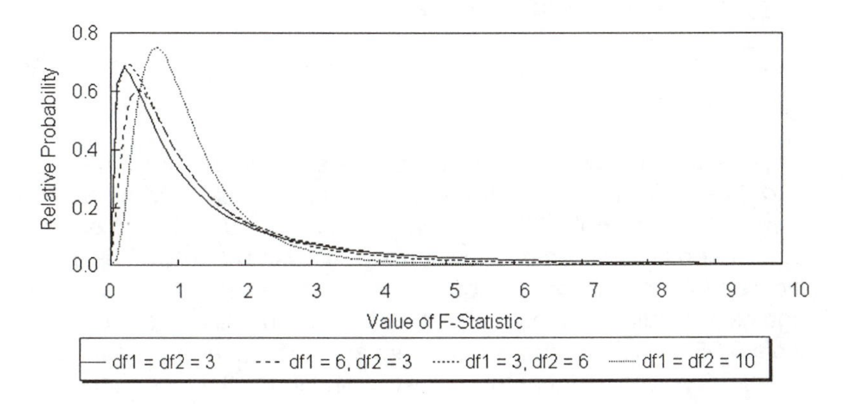

Figure A6 Form of the F-distribution for various values of the two df, df1 and df2.

A4 Tests of Significance

One of the most used tools in statistics is the test of significance, which examines the question of whether a set of data could reasonably have arisen based on a certain assumption, which is called the null hypothesis. One framework for conducting such a test has the following steps:

1. Decide on the null hypothesis to be tested (often a statement that a parameter of a distribution takes a specific value).

2. Decide whether the alternative to the null hypothesis that is of interest is that there is any difference, or whether there is a difference in a particular direction.

3. Choose a suitable test statistic which measures the extent to which the data agree with the null hypothesis.

4. Determine the distribution of the test statistic if the null hypothesis is true.

5. Calculate the test statistic, S, for the observed data.

6. Calculate the probability p (sometimes called the p-value) of obtaining a value as extreme as, or more extreme than, S if the null hypothesis is true, using the distribution identified at step (4), and defining 'extreme' taking into account the alternative to the null hypothesis identified at step (2).

7. If p is small enough, conclude that there is evidence that the null hypothesis is not true.

At step (7) it is conventional to use the probability levels 0.05, 0.01 and 0.001. Thus if $p \leq 0.05$ then the test is considered to provide some evidence against the null hypothesis, if $p \leq 0.01$ then the test is considered to provide strong evidence against the null hypothesis, and if $p \leq 0.001$ then the test is considered to provide very strong evidence against the null hypothesis. Another way to express this is to say that if $p \leq 0.05$ then the result is significant at the 5% level, if $p \leq 0.01$ then the result is significant at the 1% level, and if $p \leq 0.01$ then the result is significant at the 0.1% level. There is an element of arbitrariness in the choice of the probability levels of 0.05, 0.01 and 0.001. They were used originally because of the need to specify a limited number of levels in order to construct tables so that tests of significance could be carried out.

Some people prefer to specify in advance the level, α, of p that will be considered to be significant at step (7). For example, it might be decided that a result will be significant only if $p \leq 0.01$. The test is then said to be at the 1% level of significance, or sometimes the test is said to have a size of 0.01 or 1%. In that case the above framework changes slightly. There is an additional step before step (1):

(0) Choose the significance level α for the test;

and the last step changes to:

(7) If $p \leq \alpha$ then declare the test result to be significant, giving evidence against the null hypothesis.

Step (2) in the procedure is concerned with deciding whether the test is one-sided or two-sided. This depends on whether there is interest in differences from the null hypothesis in either direction (e.g., the true mean of a distribution could be either higher or lower than the mean specified by the null hypothesis), or only interest in differences in one direction (e.g., there is concern only if the true mean exceeds the mean specified by the null hypothesis). This then makes a difference to the calculation of the p-value at step (6). In general, what is done is to calculate the probability of a result as extreme or more extreme than that observed in terms of the direction or directions of interest. Often test statistics are such that a value of zero indicates no difference from the null hypothesis. In that case the p-value for a two-sided test will be the probability of being as far from zero as the observed value, while the p-value for a one-sided test will be the probability of being as far from zero in the direction that indicates an effect of interest. The example given at the end of this section should clarify what this means in practice.

An important distinction is between parametric and non-parametric tests. In practice this often comes down to a question of whether the population being sampled has a normal distribution or not. The difference between parametric and non-parametric tests is that parametric tests require more assumptions but use the data more efficiently. As to whether or not it is reasonable to assume a normal distribution, this depends very much on the circumstances. Some types of data are almost always normal or nearly normal. For example, measurements of lengths of body parts for animals and plants fall in this category. In other cases, a simple data transformation such as taking logarithms will change a non-normal distribution into a normal one.

The following is a brief summary of some of the most commonly used tests of significance. The book by Kanji (1999) covers just about any test that is likely to be used, with example calculations. However, in practice the easiest way to do these things is by using one of the standard statistical computer packages.

The One Sample t-Test

The purpose of this test is to see whether the mean of a random sample of n values from a population is significantly different from some hypothetical value μ. The sample mean and standard deviation ($\bar{x}$ and s) are used to calculate the statistic

$$t = (\bar{x} - \mu)/(s/\sqrt{n}).$$

As noted in Section A3, if the population mean is really μ then this follows the t-distribution with n - 1 df. Hence the significance of the sample result can be tested by seeing whether the observed statistic is a reasonable value from this distribution. This test requires the assumption that the population being sampled is normally distributed.

The One Sample Chi-Squared Test

This test is used to see whether a set of sample counts is in reasonable agreement with the frequencies expected on the basis of some hypothesis. The test statistic is

$$X^2 = \Sigma(O - E)^2/E,$$

where O is an observed frequency, E an expected frequency, and the summation is over all such frequencies. If the differences between the observed and expected frequencies are merely due to sampling errors, then the test statistic will be approximately a random value from the chi-squared distribution with n - 1 df. A significantly large value for the test statistic in comparison with the chi-squared distribution is evidence that the basis for calculating the expected frequencies is not correct. It is important to remember that the chi-squared test is only valid with count data. It is, for example, quite wrong to use this test with percentages. Also, the use of the chi-squared distribution requires that the expected frequencies are not too small, and preferably all five or more. For more details about this test see Kanji (1999, Test 37).

The Contingency Table Chi-Squared Test

A contingency table is obtained when N observations are classified on the basis of two different criteria, and a count is made of the number of observations O_{ij} that fall into the ith class for the first criterion of classification (row i of the table), and the jth class for the second criterion of classification (column j of the table). This is done for all values of i and j. For example, a 2x3 contingency table is produced if 100 streams are classified as being either in the north or the south of a region (the two row categories) and as having low, medium or high pollution (the three column categories).

The contingency table chi-squared test is designed to see whether there is any evidence that the two classifications are related, i.e., whether the probability of a sample unit being in class i for the first classification varies according to which of the classes the unit is in for the second classification. On the null hypothesis that the classifications are unrelated, it can be shown that the expected count for the cell in row i and column j of the table is $E_{ij} = R_iC_j/N$, where R_i is the total observed count in row i and C_j is the total observed count in column j. The test statistic is then

$$X^2 = \sum_i \sum_j (O_{ij} - E_{ij})^2/E_{ij},$$

where the double summation means adding over all the rows and columns in the table. If the null hypothesis of independent classifications is true, then this statistic will be approximately a random value from the chi-squared distribution with (r - 1)(c - 1) df, where r is the number of rows and c is the number of columns in the table. A significantly large value for the test statistic in comparison with the chi-squared distribution is evidence against the null hypothesis. As for the one sample chi-squared test, the expected frequencies should not be too small when this test is used. Preferably they should all be five or more. For more details about this test see Kanji (1999, Test 44).

The Paired t-Test

This is used when the data are naturally paired. The null hypothesis being tested is that the differences between paired values have a particular mean value μ_d, which is often zero. The test statistic is

$$t = (\bar{d} - \mu_d)/(s_d/\sqrt{n}),$$

where $\bar{d}$ is the mean and s_d is the standard deviation of the sample differences. This statistic is compared with the t-distribution with n - 1 df. The test requires the assumption that the paired differences are normally distributed. Essentially it is just the one sample t-test calculated on differences, with a null hypothesis mean of zero.

The Two Sample t-Test

This is designed to test whether the difference between the means of two independent samples are significantly different. The test statistic is

$$t = (\bar{x}_1 - \bar{x}_2)/\{s_p\sqrt{(1/n_1 + 1/n_2)}\},$$

where sample 1, of size n_1, has a of mean $\bar{x}_1$, and sample 2, of size n_2, has a mean $\bar{x}_2$. Also,

$$s_p^2 = \{(n_1 - 1)s_1^2 + (n_2 - 1)s_2^2\}/(n_1 + n_2 - 2)$$

is an estimate of variance based on pooling the variances, s_1^2 and s_2^2, for the two samples. To assess the significance of the sample mean difference, the calculated t-value is compared with the critical values in the t-table with $n_1 + n_2 - 2$ df. This test assumes that the two distributions being sampled are normal with the same standard deviation. For more details see Kanji (1999, Test 8). Test 9 in the same book also contains a variation on this test for situations where the distributions do not have the same standard deviation.

The Wilcoxon Signed-Ranks Test

This is a non-parametric alternative to the paired t-test. It is based on ranking the differences between two measurements and using sums of positive and negative ranks as test statistics. This test does not require normal distribution for the differences. For more details see Kanji (1999, Test 48).

The Mann-Whitney U-Test

This is a non-parametric alternative to the two sample independent t-test, which tests whether the two samples come from distributions with the same mean on the assumption that the distributions have the same shape. To use this test it is only necessary to be able to rank the data in order for the two samples combined, and no assumption of normality is required. For more details see Kanji (1999, Test 52).

Example A1 Testing the Mean Level of TcCB at a Site

As an example of a test of significance, consider the data in Table A1 on measurements of 1,2,3,4 tetrachlorobenzene (TcCB) from $n = 47$ different locations at a particular site. Suppose that a mean level of 0.5 is considered to be acceptable, and the question is whether the mean at this site is significantly higher than 0.5. A one-sample t-test is used to answer this question.

Following the framework (1) to (6) defined above then gives the following results:

1. The null hypothesis to be tested is that $\mu = 0.5$, where μ is the mean level of TcCB over the entire site.

2. The alternative of interest to the null hypothesis that needs to be detected is that μ exceeds 0.5.

3. The test statistic used is $t = (\bar{x} - 0.5)/(s/\sqrt{n})$ from equation (A4) because the data appear to follow a distribution that is reasonably close to normal (Figure A2), and this statistic measures the difference between the sample mean and the hypothetical mean of 0.5.

4. The distribution of the test statistic is a t-distribution with 46 df if the null hypothesis is true.

5. For the observed data the mean is $\bar{x} = 0.599$ and the standard deviation is $s = 0.284$. The test statistic is therefore

$$t = (0.599 - 0.5)/(0.284/\sqrt{47}) = 2.39.$$

6. Only positive values for the test statistic indicate that the mean of the site is higher than 0.5. Hence it is necessary to find the probability, p, of obtaining a mean as large as, or larger than, 2.39 for the t-distribution with 46 df. To this end, Table B2 can be consulted. This does not have a row for 46 df, but it does show that with 40 df the probability of a value as large as or larger than 2.39 is between 0.025 and 0.01, and that with 60 df the probability is 0.01. It follows that the probability of a value as large as or larger than 2.39, with 46 df, is about $p = 0.02$.

7. As the p-value is less than 0.05, it is significant at the 5% level, giving some evidence against the null hypothesis. It appears that the mean for the site is higher than 0.5.

These days, probabilities for the t-distribution are provided in spreadsheet programs. It is therefore easy enough to determine the p-value accurately for a test like this, although that is not really necessary. The exact probability of a value from the t-distribution with 46 df equalling or exceeding 2.39 is, in fact, $p = 0.0105$.

A5 Confidence Intervals

Confidence limits for a parameter of a distribution give a range within which the parameter is expected to lie. For example, 90% confidence limits for a distribution mean define a range, which is called a confidence interval, within which the mean is expected to lie 90% of the time, in the sense that if many such intervals are calculated then about 90% of them will contain the true value of the parameter.

As an example of what this means, suppose that a random sample of size 20 is taken from a normal distribution, with an unknown mean μ. Then it is known that

$$t = (\bar{x} - \mu)/(s/\sqrt{n})$$

will be a random value from the t-distribution with 19 df. For this distribution, Table B2 shows that

$$\text{Prob}(t > 1.729) = 0.05,$$

and as the distribution is symmetrical about zero, it is also true that

$$\text{Prob}(t < -1.729) = 0.05.$$

Hence, if it is asserted that

$$-1.729 < (\bar{x} - \mu)/(s/\sqrt{n}) < 1.729, \qquad \text{(A7)}$$

then this statement will be true with probability 1 - 2(0.05) = 0.9. Rearranging the left-hand side of this equation shows that

$$-1.729 < (\bar{x} - \mu)/(s/\sqrt{n})$$

is equivalent to

$$\mu < \bar{x} + 1.729(s/\sqrt{n}).$$

Similarly, rearranging the right-hand side shows that it is equivalent to

$$\bar{x} - 1.729(s/\sqrt{n}) < \mu.$$

Thus the statement (A7) is exactly equivalent to

$$\bar{x} - 1.729(s/\sqrt{n}) < \mu < \bar{x} + 1.729(s/\sqrt{n}). \qquad \text{(A8)}$$

Since statement (A7) is true with probability 0.9, statement (A8) must be true with the same probability. In this sense, statement (A8) gives a 90% confidence interval for the true population mean.

This argument can easily be generalized for a random sample of size n from a normal distribution to give a $100(1 - \alpha)\%$ confidence interval for the mean of the distribution of the form

$$\bar{x} - t_{\alpha/2,n-1}(s/\sqrt{n}) < \mu < \bar{x} + t_{\alpha/2,n-1}(s/\sqrt{n}), \qquad \text{(A9)}$$

where $t_{\alpha/2,n-1}$ is the value that is exceeded with probability $\alpha/2$ by a random value from the t-distribution with n - 1 df.

The confidence interval (A9) is one of many that can be derived for a parameter of a distribution, using a similar type of argument. A confidence interval is a range within which a parameter will lie with a certain defined probability.

Example A2 A Confidence Interval for the Mean TcCB at a Site

Considering again the data in Table A1, suppose that there is interest in determining a 95% confidence interval for the mean level of 1,2,3,4 tetrachlorobenzene at the site in question. The mean and standard deviation for the sample of size $n = 47$ are $\bar{x} = 0.599$ and $s = 0.284$, respectively, and the critical value of the t-distribution with 46 df that is exceeded with probability $\alpha/2 = 0.025$ is $t_{0.025,46} \approx 2.01$ from Table B2. Hence the 95% confidence interval (1.9) is

$$0.599 - 2.01(0.284/\sqrt{47}) < \mu < 0.599 + 2.01(0.284/\sqrt{47}),$$

i.e.,

$$0.516 < \mu < 0.682.$$

A6 Covariance and Correlation

Suppose that a random sample of size n from a large population of items is taken, and two variables X and Y are measured on each of the sampled items. Let the values of X and Y on the ith sampled item be x_i and y_i, respectively.

One sample statistic that is used to measure the relationship between X and Y is the covariance

$$c_{xy} = \sum_{i=1}^{n} (x_i - \bar{x})(y_i - \bar{y})/(n - 1), \tag{A10}$$

where $\bar{x}$ and $\bar{y}$ are the sample means for X and Y, respectively. If large values of X tend to occur with large values of Y, then c_{xy} will be positive. Conversely, if large values of X tend to occur with small values of Y, then c_{xy} will be negative.

Covariances as defined by equation (A10) are used in many statistical procedures, but have the disadvantage that there is no simple way to judge whether a particular value for c_{xy} indicates that X and Y are closely related. Therefore, for many purposes the covariance is adjusted so that it becomes the covariance between X and Y after both variables have been scaled to have sample standard deviations of exactly one. The resulting statistic is then referred to as the sample correlation, or sometimes as Pearson's correlation coefficient, which can be shown to be the same as

$$r_{xy} = c_{xy}/(s_x s_y), \tag{A11}$$

where s_x and s_y are the sample standard deviations for X and Y, as defined by equation (A2).

Values of r_{xy} range from -1 (a perfect negative linear relationship between X and Y), through 0 (no linear relationship), to +1 (a perfect positive linear relationship). Figure A7 shows plots of Y values against X values for samples of size n = 25 giving various levels for the correlation.

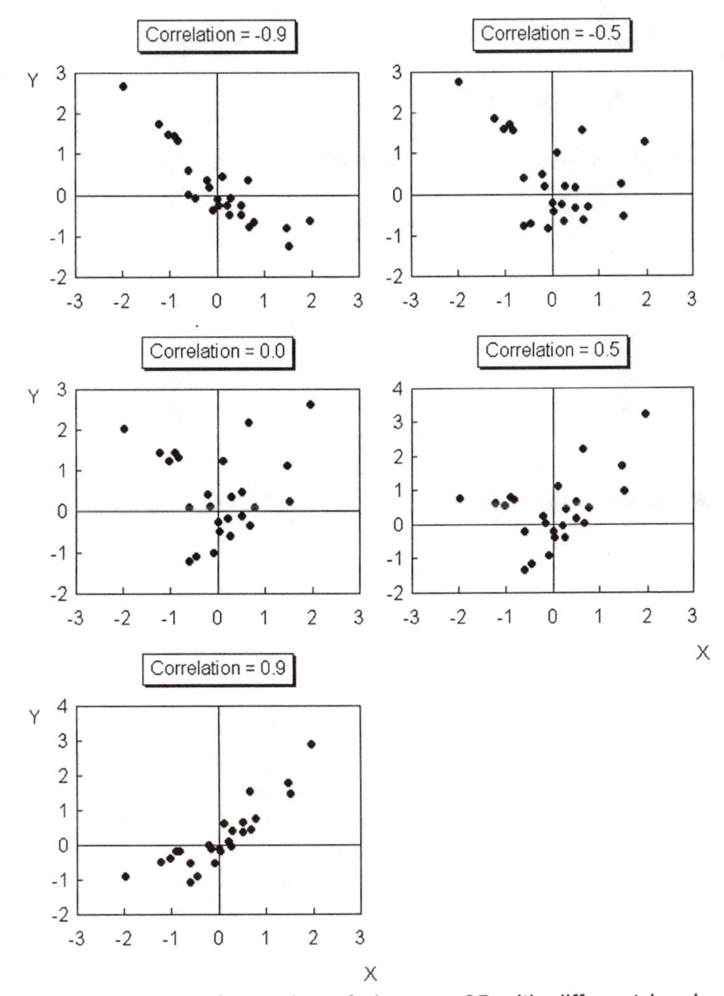

Figure A7 Examples of samples of size n = 25 with different levels of correlation between variables X and Y.

Appendix B Statistical Tables

B1 The Standard Normal Distribution

B2 Critical Values for the t-Distribution

B3 Critical Values for the Chi-Squared Distribution

B4 Critical Values for the F-Distribution

B5 Critical Values for the Durbin-Watson Statistic

B1 The Standard Normal Distribution

The entries in the table give the probability of a standard normal variable Z having a value between 0 and z. The rows of the table are for the first decimal place of z, the columns are for the second decimal place. For example, Prob(0 < Z < 1.56) = 0.441, using the row for 1.5 and the column for 0.06.

z	0.00	0.01	0.02	0.03	0.04	0.05	0.06	0.07	0.08	0.09
0.0	0.000	0.004	0.008	0.012	0.016	0.020	0.024	0.028	0.032	0.036
0.1	0.040	0.044	0.048	0.052	0.056	0.060	0.064	0.068	0.071	0.075
0.2	0.079	0.083	0.087	0.091	0.095	0.099	0.103	0.106	0.110	0.114
0.3	0.118	0.122	0.126	0.129	0.133	0.137	0.141	0.144	0.148	0.152
0.4	0.155	0.159	0.163	0.166	0.170	0.174	0.177	0.181	0.184	0.188
0.5	0.192	0.195	0.199	0.202	0.205	0.209	0.212	0.216	0.219	0.222
0.6	0.226	0.229	0.232	0.236	0.239	0.242	0.245	0.249	0.252	0.255
0.7	0.258	0.261	0.264	0.267	0.270	0.273	0.276	0.279	0.282	0.285
0.8	0.288	0.291	0.294	0.297	0.300	0.302	0.305	0.308	0.311	0.313
0.9	0.316	0.319	0.321	0.324	0.326	0.329	0.332	0.334	0.337	0.339
1.0	0.341	0.344	0.346	0.349	0.351	0.353	0.355	0.358	0.360	0.362
1.1	0.364	0.367	0.369	0.371	0.373	0.375	0.377	0.379	0.381	0.383
1.2	0.385	0.387	0.389	0.391	0.393	0.394	0.396	0.398	0.400	0.402
1.3	0.403	0.405	0.407	0.408	0.410	0.412	0.413	0.415	0.416	0.418
1.4	0.419	0.421	0.422	0.424	0.425	0.427	0.428	0.429	0.431	0.432
1.5	0.433	0.435	0.436	0.437	0.438	0.439	0.441	0.442	0.443	0.444
1.6	0.445	0.446	0.447	0.448	0.450	0.451	0.452	0.453	0.454	0.455
1.7	0.455	0.456	0.457	0.458	0.459	0.460	0.461	0.462	0.463	0.463
1.8	0.464	0.465	0.466	0.466	0.467	0.468	0.469	0.469	0.470	0.471
1.9	0.471	0.472	0.473	0.473	0.474	0.474	0.475	0.476	0.476	0.477
2.0	0.477	0.478	0.478	0.479	0.479	0.480	0.480	0.481	0.481	0.482
2.1	0.482	0.483	0.483	0.483	0.484	0.484	0.485	0.485	0.485	0.486
2.2	0.486	0.486	0.487	0.487	0.488	0.488	0.488	0.488	0.489	0.489
2.3	0.489	0.490	0.490	0.490	0.490	0.491	0.491	0.491	0.491	0.492
2.4	0.492	0.492	0.492	0.493	0.493	0.493	0.493	0.493	0.493	0.494
2.5	0.494	0.494	0.494	0.494	0.495	0.495	0.495	0.495	0.495	0.495
2.6	0.495	0.496	0.496	0.496	0.496	0.496	0.496	0.496	0.496	0.496
2.7	0.497	0.497	0.497	0.497	0.497	0.497	0.497	0.497	0.497	0.497
2.8	0.497	0.498	0.498	0.498	0.498	0.498	0.498	0.498	0.498	0.498
2.9	0.498	0.498	0.498	0.498	0.498	0.498	0.499	0.499	0.499	0.499
3.0	0.499	0.499	0.499	0.499	0.499	0.499	0.499	0.499	0.499	0.499

B2 Critical Values for the t-Distribution

The entries in the table give the critical values that are exceeded with probabilities of 0.05, 0.025, 0.01 and 0.005 by random values from t-distributions with different numbers of degrees of freedom (df). For example, if T has a t-distribution with 12 df, then Prob(T > 2.179) = 0.025.

	Upper tail probability			
df	0.050	0.025	0.010	0.005
1	6.314	12.706	31.821	63.657
2	2.920	4.303	6.965	9.925
3	2.353	3.182	4.541	5.841
4	2.132	2.776	3.747	4.604
5	2.015	2.571	3.365	4.032
6	1.943	2.447	3.143	3.708
7	1.895	2.365	2.998	3.500
8	1.860	2.306	2.897	3.355
9	1.833	2.262	2.821	3.250
10	1.812	2.228	2.764	3.169
11	1.796	2.201	2.718	3.106
12	1.782	2.179	2.681	3.055
13	1.771	2.160	2.650	3.012
14	1.761	2.145	2.625	2.977
15	1.753	2.132	2.603	2.947
16	1.746	2.120	2.584	2.921
17	1.740	2.110	2.567	2.898
18	1.734	2.101	2.552	2.878
19	1.729	2.093	2.540	2.861
20	1.725	2.086	2.528	2.845
21	1.721	2.080	2.518	2.831
22	1.717	2.074	2.508	2.819
23	1.714	2.069	2.500	2.807
24	1.711	2.064	2.492	2.797
25	1.708	2.060	2.485	2.787
26	1.706	2.056	2.479	2.779
27	1.703	2.052	2.473	2.771
28	1.701	2.048	2.467	2.763
29	1.699	2.045	2.462	2.756
30	1.697	2.042	2.457	2.750
40	1.684	2.021	2.423	2.705
60	1.671	2.000	2.390	2.660
120	1.658	1.980	2.358	2.617
Inf	1.645	1.960	2.326	2.576

B3 Critical Values for the Chi-Squared Distribution

The entries in the table give the critical values that are exceeded with probabilities of 0.1, 0.05, 0.025, 0.01, 0.005, and 0.001 by random values from chi-squared distributions with different numbers of degrees of freedom (df). For example, if X^2 has 4 df, then $\text{Prob}(X^2 > 9.49) = 0.05$.

	Upper tail probability					
df	0.1	0.05	0.025	0.01	0.005	0.001
1	2.71	3.84	5.02	6.64	7.88	10.83
2	4.61	5.99	7.38	9.21	10.60	13.82
3	6.25	7.82	9.35	11.35	12.84	16.27
4	7.78	9.49	11.14	13.28	14.86	18.47
5	9.24	11.07	12.83	15.09	16.75	20.52
6	10.65	12.59	14.45	16.81	18.55	22.46
7	12.02	14.07	16.01	18.48	20.28	24.32
8	13.36	15.51	17.54	20.09	21.96	26.13
9	14.68	16.92	19.02	21.67	23.59	27.88
10	15.99	18.31	20.48	23.21	25.19	29.59
11	17.28	19.68	21.92	24.73	26.76	31.26
12	18.55	21.03	23.34	26.22	28.30	32.91
13	19.81	22.36	24.74	27.69	29.82	34.53
14	21.06	23.69	26.12	29.14	31.32	36.12
15	22.31	25.00	27.49	30.58	32.80	37.70
16	23.54	26.30	28.85	32.00	34.27	39.25
17	24.77	27.59	30.19	33.41	35.72	40.79
18	25.99	28.87	31.53	34.81	37.16	42.31
19	27.20	30.14	32.85	36.19	38.58	43.82
20	28.41	31.41	34.17	37.57	40.00	45.32
21	29.62	32.67	35.48	38.93	41.40	46.80
22	30.81	33.92	36.78	40.29	42.80	48.27
23	32.01	35.17	38.08	41.64	44.18	49.73
24	33.20	36.42	39.36	42.98	45.56	51.18
25	34.38	37.65	40.65	44.31	46.93	52.62
26	35.56	38.89	41.92	45.64	48.29	54.05
27	36.74	40.11	43.20	46.96	49.65	55.48
28	37.92	41.34	44.46	48.28	50.99	56.89
29	39.09	42.56	45.72	49.59	52.34	58.30
30	40.26	43.77	46.98	50.89	53.67	59.70
40	51.81	55.76	59.34	63.69	66.77	73.41
50	63.17	67.51	71.42	76.15	79.49	86.66
60	74.40	79.08	83.30	88.38	91.96	99.62
70	85.53	90.53	95.02	100.42	104.21	112.31
80	96.58	101.88	106.63	112.33	116.32	124.84
90	107.57	113.15	118.14	124.12	128.30	137.19
100	118.50	124.34	129.56	135.81	140.18	149.48

B4 Critical Values for the F-Distribution

The entries in the table give the critical values that are exceeded with probability 0.05 by random values from F-distributions with v_1 degrees of freedom (df) in the numerator and v_2 df in the denominator. For example, if F has $v_1 = 10$ and $v_2 = 4$ df, then Prob($F > 5.96$) = 0.05.

	v_1														
v_2	1	2	3	4	5	6	7	8	9	10	12	15	20	30	60
1	161.5	199.5	215.7	224.6	230.2	234.0	236.8	238.9	240.5	241.9	243.9	246.0	248.0	250.1	252.2
2	18.51	19.00	19.16	19.25	19.30	19.33	19.35	19.37	19.39	19.40	19.41	19.43	19.45	19.46	19.48
3	10.13	9.55	9.28	9.12	9.01	8.94	8.89	8.85	8.81	8.79	8.75	8.70	8.66	8.62	8.57
4	7.71	6.94	6.59	6.39	6.26	6.16	6.09	6.04	6.00	5.96	5.91	5.86	5.80	5.75	5.69
5	6.61	5.79	5.41	5.19	5.05	4.95	4.88	4.82	4.77	4.74	4.68	4.62	4.56	4.50	4.43
6	5.99	5.14	4.76	4.53	4.39	4.28	4.21	4.15	4.10	4.06	4.00	3.94	3.87	3.81	3.74
7	5.59	4.74	4.35	4.12	3.97	3.87	3.79	3.73	3.68	3.64	3.58	3.51	3.45	3.38	3.30
8	5.32	4.46	4.07	3.84	3.69	3.58	3.50	3.44	3.39	3.35	3.28	3.22	3.15	3.08	3.01
9	5.12	4.26	3.86	3.63	3.48	3.37	3.29	3.23	3.18	3.14	3.07	3.01	2.94	2.86	2.79
10	4.97	4.10	3.71	3.48	3.33	3.22	3.14	3.07	3.02	2.98	2.91	2.85	2.77	2.70	2.62
11	4.84	3.98	3.59	3.36	3.20	3.10	3.01	2.95	2.90	2.85	2.79	2.72	2.65	2.57	2.49
12	4.75	3.89	3.49	3.26	3.11	3.00	2.91	2.85	2.80	2.75	2.69	2.62	2.54	2.47	2.38
13	4.67	3.81	3.41	3.18	3.03	2.92	2.83	2.77	2.71	2.67	2.60	2.53	2.46	2.38	2.30
14	4.60	3.74	3.34	3.11	2.96	2.85	2.76	2.70	2.65	2.60	2.53	2.46	2.39	2.31	2.22
15	4.54	3.68	3.29	3.06	2.90	2.79	2.71	2.64	2.59	2.54	2.48	2.40	2.33	2.25	2.16
16	4.49	3.63	3.24	3.01	2.85	2.74	2.66	2.59	2.54	2.49	2.43	2.35	2.28	2.19	2.11
17	4.45	3.59	3.20	2.97	2.81	2.70	2.61	2.55	2.49	2.45	2.38	2.31	2.23	2.15	2.06
18	4.41	3.56	3.16	2.93	2.77	2.66	2.58	2.51	2.46	2.41	2.34	2.27	2.19	2.11	2.02
19	4.38	3.52	3.13	2.90	2.74	2.63	2.54	2.48	2.42	2.38	2.31	2.23	2.16	2.07	1.98
20	4.35	3.49	3.10	2.87	2.71	2.60	2.51	2.45	2.39	2.35	2.28	2.20	2.12	2.04	1.95

B4 Critical Values for the F-Distribution (Continued)

v_2	v_1 = 1	2	3	4	5	6	7	8	9	10	12	15	20	30	60
21	4.33	3.47	3.07	2.84	2.69	2.57	2.49	2.42	2.37	2.32	2.25	2.18	2.10	2.01	1.92
22	4.30	3.44	3.05	2.82	2.66	2.55	2.46	2.40	2.34	2.30	2.23	2.15	2.07	1.98	1.89
23	4.28	3.42	3.03	2.80	2.64	2.53	2.44	2.38	2.32	2.28	2.20	2.13	2.05	1.96	1.87
24	4.26	3.40	3.01	2.78	2.62	2.51	2.42	2.36	2.30	2.26	2.18	2.11	2.03	1.94	1.84
25	4.24	3.39	2.99	2.76	2.60	2.49	2.41	2.34	2.28	2.24	2.17	2.09	2.01	1.92	1.82
26	4.23	3.37	2.98	2.74	2.59	2.47	2.39	2.32	2.27	2.22	2.15	2.07	1.99	1.90	1.80
27	4.21	3.35	2.96	2.73	2.57	2.46	2.37	2.31	2.25	2.20	2.13	2.06	1.97	1.88	1.79
28	4.20	3.34	2.95	2.71	2.56	2.45	2.36	2.29	2.24	2.19	2.12	2.04	1.96	1.87	1.77
29	4.18	3.33	2.93	2.70	2.55	2.43	2.35	2.28	2.22	2.18	2.11	2.03	1.95	1.85	1.75
30	4.17	3.32	2.92	2.69	2.53	2.42	2.33	2.27	2.21	2.17	2.09	2.02	1.93	1.84	1.74
40	4.09	3.23	2.84	2.61	2.45	2.34	2.25	2.18	2.12	2.08	2.00	1.92	1.84	1.74	1.64
60	4.00	3.15	2.76	2.53	2.37	2.25	2.17	2.10	2.04	1.99	1.92	1.84	1.75	1.65	1.53
120	3.92	3.07	2.68	2.45	2.29	2.18	2.09	2.02	1.96	1.91	1.83	1.75	1.66	1.55	1.43

B5 Critical Values for the Durbin-Watson Statistic

This table gives bounds for critical values for the Durbin-Watson statistic V, for a two-sided test (for positive or negative autocorrelation) at the 5% level (n = number of data points, p = number of X variables in the regression). If V is less than 2, then it is definitely significant if $V < d_1$, may be significant if $d_1 < V < d_2$, and is not significant if $V > d_2$. If V is greater than 2, then it is definitely significant if $4 - V < d_1$, may be significant if $d_1 < 4 - V < d_2$, and not significant if $4 - V > d_2$. For example, if there are n = 30 observations and p = 2 X variables, then to be definitely significant V must be less than 1.18, or 4 - V must be less than 1.18.

	p = 1		p = 2		p = 3		p = 4		p = 5	
n	d_1	d_2	d_1	d_2	d_1	d_2	d_1	d_2	d_1	d_2
15	0.95	1.22	0.83	1.41	0.71	1.61	0.59	1.84	0.48	2.09
20	1.08	1.29	0.99	1.41	0.89	1.55	0.79	1.70	0.70	1.87
25	1.18	1.34	1.10	1.43	1.02	1.54	0.94	1.65	0.86	1.77
30	1.25	1.38	1.18	1.46	1.12	1.54	1.05	1.63	0.98	1.73
35	1.31	1.42	1.25	1.48	1.19	1.55	1.13	1.63	1.07	1.70
40	1.35	1.45	1.30	1.51	1.25	1.57	1.20	1.63	1.15	1.69
45	1.39	1.48	1.34	1.53	1.30	1.58	1.25	1.63	1.21	1.69
50	1.42	1.50	1.38	1.54	1.34	1.59	1.30	1.64	1.26	1.69
60	1.47	1.54	1.44	1.57	1.40	1.61	1.37	1.65	1.33	1.69
70	1.51	1.57	1.48	1.60	1.45	1.63	1.42	1.66	1.39	1.70
80	1.54	1.59	1.52	1.62	1.49	1.65	1.47	1.67	1.44	1.70
90	1.57	1.61	1.55	1.64	1.53	1.66	1.50	1.69	1.48	1.71
100	1.59	1.63	1.57	1.65	1.55	1.67	1.53	1.70	1.51	1.72

References

Akritas, M.G., Ruscitti, T.F. and Patil, G.P. (1994). Statistical analysis of censored environmental data. In *Handbook of Statistics 12: Environmental Statistics* (Eds. G.P. Patil and C.R. Rao), pp. 221-42. North-Holland, Amsterdam.

Andersen, M. (1992). Spatial analysis of two-species interactions. *Ecology* 91: 134-40.

Association for the Environmental Health of Soils (2000). *Human and Ecological Risk Assessment: Content of Volume 2, Number 4*. Available from the web site www.aehs.com/.

Baird, S.J. (1996). Nonfish Species and Fisheries Interactions Working Group Report, May 1996. New Zealand Fisheries Assessment Working Group Report 96/1. Ministry of Fisheries, Wellington, New Zealand.

Barnett, V. and Turkman, K.F., Eds. (1993). *Statistics for the Environment*. Wiley, Chichester.

Barnett, V. and Turkman, K.F., Eds. (1994). *Statistics for the Environment 2: Water Related Issues*. Wiley, Chichester.

Barnett, V. and Turkman, K.F., Eds. (1997). *Statistics for the Environment 3: Pollution Assessment and Control*. Wiley, Chichester.

Berger, R.L. and Hsu, J.C. (1996). Bioequivalence trials, intersection-union tests and equivalence confidence sets. *Statistical Science* 11: 283-319.

Bergman, H.L., Meyer, J.S., Marr, J.C.A., Hansen, J.A., Szumski, M.J., Farag, A.M., MacRae, R.K., Parrish, T.L., Hill, S.L., Boelter, A.M., McDonald, L., Johnson, G., Strickland, D., Dean, T. and Rowe, R. (1995). *Guidance Document for Determination of Injury to Biological Resources Resulting from Incidents Involving Oil*. National Oceanic and Atmospheric Administration, Silver Spring, Maryland.

Besag, J. (1978). Some methods of statistical analysis for spatial pattern. *Bulletin of the International Statistical Institute* 47: 77-92.

Besag, J. and Diggle, P.J. (1977). Simple Monte Carlo tests for spatial pattern. *Applied Statistics* 26: 327-33.

Borgman, L., Taheri, M. and Hagan, R. (1984). Three-dimensional frequency-domain simulations of geologic variables. In *Geostatistics for Natural Resources Characterizations* (Eds. G. Verley, A.G. Journel, and A. Marechal), pp. 517-41. Reidel, Dordrecht.

Borgman, L.E., Gerow, K. and Flatman, G.T. (1996). Cost-effective sampling of spatially distributed phenomena. In *Principles of Environmental Sampling*, 2nd Edit. (Ed. L.H. Keith), pp. 753-78. American Chemical Society, Washington, DC.

Bradley, J.V. (1968). *Distribution Free Statistical Methods*. Prentice-Hall, New Jersey.

Burnham, K.P., Anderson, D.R., White, G.C., Brownie, C. and Pollock, K.H. (1987). *Design and Analysis of Fish Survival Experiments Based on Release-Recapture*. American Fisheries Society Monograph 5, Bethesda, Maryland.

Camacho-Ibar, V.F. and McEvoy, J. (1996). Total PCBs in Liverpool Bay Sediment. *Marine Environmental Research* 41: 241-63.

Campbell, D.T. and Stanley, J.C. (1963). *Experimental and Quasi-Experimental Designs for Research*. Houghton Mifflin, Boston.

Carlin, B.P., Gelfand, A.E. and Smith, A.F.M. (1992). Hierarchical Bayesian analysis of changepoint problems. *Applied Statistics* 41: 389-405.

Carpenter, S.R., Chisholm, S.W., Krebs, C.J., Schindler, D.W. and Wright, R.F. (1995). Ecosystem Experiments. *Science* 269: 324-7.

Carpenter, S.R., Frost, T.M., Heisey, D. and Kratz, T.K. (1989). Randomized intervention analysis and the interpretation of whole-ecosystem experiments. *Ecology* 70: 1142-52.

Caselton, W.F., Kan, L. and Zidek, J.V. (1992). Quality data networks that minimize entropy. In *Statistics in the Environmental and Earth Sciences* (Eds. A.T. Walden and P. Guttorp), pp. 10-38. Edward Arnold, London.

CCAMLR (1992). *Basic Documents*, 6th Edition. Committee for the Conservation of Antarctic Marine Living Resources, Hobart, Australia.

Chatfield, C. (1989). *The Analysis of Time Series: an Introduction*, 4th edit. Chapman and Hall, London.

Cherry, S. (1998). Statistical tests in publications of the Wildlife Society. *Wildlife Society Bulletin* 26: 947-53.

Clarke, J.U. (1994). Evaluating methods for statistical analysis of less than detection limit data using simulated small samples. 2. General results. In *Proceedings of the Second International Conference on Dredging and Dredged Material Placement* (Ed., E.C. McNair), pp. 747-55. American Society of Civil Engineers, New York.

Cochran, W.G. (1977). *Sampling Techniques*, 3rd edit. Wiley, New York.

Cohen, A.C. (1959). Simplified estimators for the normal distribution when samples are singly censored or truncated. *Technometrics* 1: 217-37.

Cohen, A.C. (1991). *Truncated and Censored Samples: Theory and Applications*. Marcel Dekker, New York.

Conquest, L.L. and Ralph, S.C. (1998). Statistical design considerations for monitoring and assessment. In *River Ecology and Management: Lessons from the Pacific Coastal Ecoregion* (Eds. R.J. Naiman and R.E. Bilby), pp. 455-75. Springer-Verlag, New York.

Cormack, R.M. (1994). Statistical thoughts on the UK environmental change network. In *Statistics in Ecology and Environmental Monitoring* (Eds. D.J. Fletcher and B.F.J. Manly), pp. 159-72. University of Otago Press, Dunedin.

Dale, M.R.T. and MacIsaac, D.A. (1989). New methods for the analysis of spatial pattern in vegetation. *Journal of Ecology* 77: 78-91.

Dauble, D.D., Skalski, J., Hoffman, A. and Giorgi, A.E. (1993). *Evaluation and Application of Statistical Methods for Estimating Smolt Survival*. Bonneville Power Administration, P.O. Box 3621, Portland, Oregon.

David, M. (1977). *Geostatistical Ore Reserve Estimation*. Elsevier, Amsterdam.

Davies, O.L. and Goldsmith, P.L., Eds. (1972). *Statistical Methods in Research and Production*. Oliver and Boyd, Edinburgh.

Decisioneering Inc. (2000). Crystal Ball 2000 spreadsheet add-on for Monte Carlo simulation. Information available from the web site www.decisioneering.com/crystal_ball/index.html.

Dennis, B. (1996). Discussion: should ecologists become Bayesians? *Ecological Applications* 6: 1095-103.

Deutsch, C.V. and Journel, A.G. (1992). *GSLIB, Geostatistical Software Library and User's Guide*. Oxford University Press, New York.

Dixon, P. (1998). Testing for no effect when the data contain below-detection values. In *Statistics in Ecology and Environmental Monitoring* (Eds. D.J. Fletcher and B.F.J. Manly), pp. 17-32. University of Otago Press, Dunedin.

Dominici, F., Parmigiani, G., Reckhow, K.H. and Wolpert, R.L. (1997). Combining information from related regressions. *Journal of Agricultural, Biological and Environmental Statistics* 2: 313-32.

Durbin, J. and Watson, G.S. (1951). Testing for serial correlation in least squares regression. *Biometrika* 38: 159-78.

Eberhardt, L.L. and Thomas, J.M. (1991). Designing environmental field studies. *Ecological Monographs* 61: 53-73.

Edgington, E.S. (1987). *Randomization Tests*, 2nd edit. Marcel Dekker, New York.

Edwards, D. and Coull, B.C. (1987). Autoregressive trend analysis: an example using long-term ecological data. *Oikos* 50: 95-102.

Efron, B. (1979). Bootstrap methods - another look at the jackknife. *Annals of Statistics* 7: 1-26.

Efron, B. (1981). Nonparametric standard errors and confidence intervals. *Canadian Journal of Statistics* 9: 139-72.

Efron, B. and Tibshirani, R.J. (1993). *An Introduction to the Bootstrap*. Chapman and Hall, New York.

El-Shaarawi, A.H. and Piegorsch, W.W., Chief Eds. (2001). *Encyclopedia of Environmetrics*, Wiley, Chichester.

Fedorov, V. and Mueller, W. (1989). Comparison of two approaches in the optimal design of an observation network. *Statistics* 20: 339-51.

Finley, B.L., Scott, P. and Paustenbach, D.J. (1993). Evaluating the adequacy of maximum contaminant levels as health-protective cleanup goals: an analysis based on Monte Carlo techniques. *Regulatory Toxicology and Pharmacology* 18: 438- 55.

Finney, D.J. (1941). On the distribution of a variate whose logarithm is normally distributed. *Journal of the Royal Statistical Society* B7: 155-61.

Fisher, R.A. (1935). *The Design of Experiments*. Oliver and Boyd, Edinburgh.

Fisher, R.A. (1936). The coefficient of racial likeness and the future of craniometry. *Journal of the Royal Anthropological Institute* 66: 57-63.

Fisher, R.A. (1970). *Statistical Methods for Research Workers*, 14th edit. Oliver and Boyd, Edinburgh.

Fletcher, D.J. and Manly, B.F.J., Eds. (1994). *Statistics in Ecology and Environmental Monitoring*. University of Otago Press, Dunedin.

Fletcher, D.J., Kavalieris, L. and Manly, B.F.J., Eds. (1998). *Statistics in Ecology and Environmental Monitoring 2: Decision Making and Risk Assessment in Biology*. University of Otago Press, Dunedin.

Folks, J.L. (1984). Combination of independent tests. In *Handbook of Statistics 4, Nonparametric Methods* (Eds. P.R. Krishnaiah and P.K. Sen), pp. 113-21. North-Holland, Amsterdam.

Francis, B., Green, M. and Payne, C. (1993). *The GLIM System, Release 4 Manual.* Clarendon Press, Oxford.

Fritts, H.C. (1976). *Tree-Rings and Climate*. Academic Press, London.

Galiano, E.F., Castro, I. and Sterling, A. (1987). A test for spatial pattern in vegetation using a Monte Carlo simulation. *Journal of Ecology* 75: 915-24.

Gardner, M.J. and Altman, D.G. (1986). Confidence intervals rather than p-values: estimation rather than hypothesis testing. *British Medical Journal* 292: 746-50.

Gelman, A., Carlin, J.B., Stern, H.S. and Rubin, D.R. (1995). *Bayesian Data Analysis*. Chapman and Hall, London.

Gibbons, J.D. (1986). Randomness, tests of. *Encyclopedia of Statistical Sciences* 7: 555-62. Wiley, New York.

Gilbert, R.O. (1987). *Statistical Methods for Environmental Pollution Monitoring*. Van Nostrand Reinhold, New York.

Gilbert, R.O. and Simpson, J.C. (1992). *Statistical Methods for Evaluating the Attainment of Cleanup Standards. Volume 3: Reference-Based Standards for Soils and Solid Media.* United States Environmental Protection Agency Report

PNL-7409, National Technical Information Service, Springfield, Virginia 22161, USA.

Gilfillan, E.S., Page, D.S., Harner, E.J. and Boehm, P.D. (1995). Shoreline Ecology Program for Prince William Sound, Alaska, following the *Exxon Valdez* oil spill: Part 3 - biology. In *Exxon Valdez Oil Spill: Fate and Effects in Alaskan Waters* (Eds. P.G. Wells, J.N. Butler and J.S. Hughes), pp. 398-441. American Society for Testing and Materials, Philadelphia.

Gilks, W.R., Richardson, S. and Spiegelhalter, D.J., Eds. (1996). *Markov Chain Monte Carlo in Practice*. Chapman and Hall, London.

Gilliom, R. and Helsel, D. (1986). Estimation of distributional parameters for censored trace level water quality data. 1. Estimation techniques. *Water Resources Research* 22: 135-46.

Gleit, A. (1985). Estimation for small normal data sets with detection limits. *Environmental Science and Technology* 19: 1201-6.

Gonzalez, L. and Benwell, G.L. (1994). Stochastic models of the behaviour of scrub weeds in Southland and Otago. In *Statistics in Ecology and Environmental Monitoring* (Eds. D.J. Fletcher and B.F.J. Manly), pp. 111-23. University of Otago Press, Dunedin.

Good, P. (1994). *Permutation Tests: A Practical Guide to Resampling Methods for Testing Hypotheses*. Springer-Verlag, New York.

Goovaerts, P. (1997). *Geostatistics for Natural Resource Evaluation*. Oxford University Press, New York.

Gore, S.D. and Patil, G.P. (1994). Identifying extremely large values using composite sample data. *Environmental and Ecological Statistics* 1: 227-45.

Gotelli, N.J. and Graves, G.R. (1996). *Null Models in Ecology*. Smithsonian Institution Press, Washington, DC.

Green, E.L. (1973). Location analysis of prehistoric Maya in British Honduras. *American Antiquity* 38: 279-93.

Green, R.H. (1979). *Sampling Design and Statistical Methods for Environmental Biologists*. Wiley, New York.

Gurevitch, J. and Hedges, L.V. (1993). Meta-analysis: combining the results of independent studies in experimental ecology. In *The Design and Analysis of Ecological Experiments* (Eds. S. Scheiner and J. Gurevitch), pp. 378-98. Chapman and Hall, New York.

Gurevitch, J. and Hedges, L.V. (1999). Statistical issues in ecological meta-analysis. *Ecology* 80: 1142-49.

Haase, P. (1995). Spatial pattern analysis in ecology based on Ripley's K-function: introduction and method of edge correction. *Journal of Vegetation Science* 6: 575-82.

Hairston, N.G. (1989). *Ecological Experiments: Purpose, Design, and Execution*. Cambridge University Press, Cambridge.

Hall, P. and Wilson, S. (1991). Two guidelines for bootstrap hypothesis testing. *Biometrics* 47: 757-62.

Harcum, J.B., Loftis, J.C. and Ward, R.C. (1992). Selecting trend tests for water quality series with serial correlation and missing values. *Water Resources Bulletin* 28: 469-78.

Harkness, R.D. and Isham, V. (1983). A bivariate spatial point pattern of ants' nests. *Applied Statistics* 32: 293-303.

Harner, E.J., Gilfillan, E.S. and O'Reilly, J.E. (1995). A comparison of the design and analysis strategies used in assessing the ecological consequences of the *Exxon Valdez*. Paper presented at the International Environmetrics Conference, Kuala Lumpur, December 1995.

Healy, M.J.R. (1988). *GLIM: An Introduction*. Clarenden Press, Oxford.

Hedges, L.V. and Olkin, I. (1985). *Statistical Methods for Meta-Analysis*. Academic Press, New York.

Hedges, L.V. and Olkin, I. (1999). *Statistical Methods for Meta-Analysis in the Medical and Social Sciences*. Academic Press, New York.

Helsel, D. and Cohn, T. (1988). Estimation of descriptive statistics for multiple censored water quality data. *Water Resources Research* 24: 1997-2004.

Helsel, D.R. and Hirsch, R.M. (1992). *Statistical Methods in Water Resources*. Elsevier, Amsterdam.

Highsmith, R.C., Stekoll, M.S., Barber, W.E., Deysher, L., McDonald, L., Strickland, D. and Erickson, W.P. (1993). *Comprehensive Assessment of Coastal Habitat, Final Status Report, Vol. I, Coastal Habitat Study No. 1A.* School of Fisheries and Ocean Sciences, University of Fairbanks, Alaska.

Hintze, J. (1998). *NCSS 2000.* Number Cruncher Statistical Systems, 329 North 1000 East, Kaysville, Utah 84037, USA.

Hirsch, R.M. and Slack, J.R. (1984). A nonparametric trend test for seasonal data with serial dependence. *Water Resources Research* 20: 727-32.

Hirsch, R.M., Slack, J.R. and Smith, R.A. (1982). Techniques of trend analysis for monthly water quality data. *Water Resources Research* 18: 107-21.

Hochberg, Y. and Tamhane, A.C. (1987). *Multiple Comparison Procedures*. Wiley, New York.

Holm, S. (1979). A simple sequential rejective multiple test procedure. *Scandinavian Journal of Statistics* 6: 65-70.

Holyoak, M. and Crowley, P.H. (1993). Avoiding erroneously high levels of detection in combinations of semi-independent tests. *Oecologia* 95: 103-14.

Horvitz, D.G. and Thompson, D.J. (1952). A generalization of sampling without replacement from a finite universe. *Journal of the American Statistical Association* 47: 663-85.

Houghton, J.P., Lees, D.C. and Driskell, W.B. (1993). *Evaluation of the Condition of Prince William Sound Shoreline Following the Exxon Valdez Oil Spill and Subsequent Shoreline Treatment, Vol. II: 1992 Biological Monitoring Survey.* National Oceanic and Atmospheric Administration Technical Memorandum NOS ORCA 73, Seattle, Washington.

Hurlbert, S. H. (1984). Pseudoreplication and the design of ecological field experiments. *Ecological Monographs* 54:187-211.

Iwamoto, R.N., Muir, W.D., Sandford, B.P., McIntyre, K.W., Frost, D.A., Williams, J.G., Smith, S.G. and Skalski, J.R. (1994). *Survival Estimates for the Passage of Juvenile Chinook Salmon through Snake River Dams and Reservoirs: Annual Report 1993.* Bonneville Power Administration, P.O. Box 3621, Portland, Oregon.

Jandhyala, V.K., Fotopoulus, S.B. and Evaggelopoulos, N. (1999). Change-point methods for Weibull models with applications to detection of trends in extreme temperatures. *Environmetrics* 10: 547-64.

Jandhyala, V.K. and MacNeill, I.B. (1986). The change point problem: a review of applications. In *Statistical Aspects of Water Quality Monitoring* (Eds. A.H. El-Shaarawi and R.E. Kwiatkowski), pp. 381-7. Elsevier, Amsterdam.

Johnson, D.H. (1999). The insignificance of significance testing. *Journal of Wildlife Management* 63: 763-72.

Johnson, N.L. and Kotz, S. (1969). *Discrete Distributions*. Wiley, New York.

Johnson, N.L. and Kotz, S. (1970a). *Continuous Univariate Distributions - 1*. Wiley, New York.

Johnson, N.L. and Kotz, S. (1970b). *Continuous Univariate Distributions - 2*. Wiley, New York.

Jones, P.D., Briffa, K.R., Barnett, T.P. and Tett, S.F.B. (1998a). High resolution palaeoclimatic records for the last millenium: interpretation, integration and comparison with general circulation model control-run temperatures. *The Holocene* 8: 455-71.

Jones, P.D., Briffa, K.R., Barnett, T.P. and Tett, S.F.B. (1998b). Millenial temperature reconstructions. IGBP Pages/World Data Center-A for Paleoclimatology Data Contribution Series #1998-039. NOAA/NGDC Paleoclimatology Program, Boulder, Colorado.

Journel, A.G. and Huijbregts, C.J. (1978). *Mining Geostatistics*. Academic Press, New York.

Judge, G.G., Hill, R.C., Griffiths, W.E., Lutkepohl, H. and Lee, T. (1988). *Introduction to the Theory and Practice of Econometrics*. Wiley, New York.

Kanji, G.K. (1999). *100 Statistical Tests*, New edit. Sage Publications, London.

Keith, L.H. (1991). *Environmental Sampling and Analysis: A Practical Guide*. Lewis Publishers, Chelsea, Michigan.

Keith, L.H., Ed. (1996). *Principles of Environmental Sampling*, 2nd edit. American Chemical Society, Washington, DC.

Kirkwood, T. B. L. (1981). Bioequivalence testing-a need to rethink. *Biometrics* 37: 589-94

Krige, D.G. (1966). Two dimensional weighted moving average trend surfaces for ore evaluation. *Journal of the South African Institute for Mining and Metallurgy* 66: 13-38.

Lambert, D., Peterson, B. and Terpenning, I. (1991). Nondetects, detection limits, and the probability of detection. *Journal of the American Statistical Association* 86: 266-77.

Leadbetter, M.R., Lindgren, G. and Rootzen, H. (1983). *Extremes and Related Properties of Random Sequences and Series*. Springer-Verlag, New York.

Lemeshow, S., Hosmer, D.W., Klar, J. and Lwanga, S.K. (1990). *Adequacy of Sample Size in Health Studies*. Wiley, Chichester.

Lesser, V.M. and Kalsbeek, W.D. (1997). A comparison of periodic survey designs employing multi-stage sampling. *Environmental and Ecological Statistics* 4: 117-30.

Liabastre, A.A., Carlesberg, K.A. and Miller, M.S. (1992). Quality assurance for environmental assessment activities. In *Methods of Environmental Data Analysis* (Ed. C.N. Hewitt), pp. 259-99. Chapman and Hall, London.

Liebhold, A. and Sharov, A. (1997). Analysis of insect count data: testing for correlation in the presence of autocorrelation. In *Population and Community Ecology for Insect Management and Conservation* (Eds. J. Baumgärtner, P. Bradmayr and B.F.J. Manly), pp. 111-7. Balkema, Rotterdam.

Lindsey, J.K. (1989). *The Analysis of Categorical Data Using GLIM*. Springer-Verlag, Berlin.

Loftis, J.C., McBride, G.B. and Ellis, J.C. (1991). Considerations of scale in water quality monitoring and data analysis. *Water Resources Bulletin* 27: 255-64.

Lotwick, H.W. and Silverman, B.W. (1982). Methods for analysing spatial processes of several types of point. *Journal of the Royal Statistical Society* B44: 406-13.

MacNally, R. and Hart, B.T. (1997). Use of CUSUM methods for water quality monitoring in storages. *Environmental Science and Technology* 31: 2114-9.

Madansky, A. (1988). *Prescriptions for Working Statisticians*. Springer-Verlag, New York.

Manly, B.F.J. (1992). *The Design and Analysis of Research Studies*. Cambridge University Press, Cambridge.

Manly, B.F.J. (1994). CUSUM methods for detecting changes in monitored environmental variables. In *Statistics in Ecology and Environmental Monitoring* (Eds. D.J. Fletcher and B.F.J. Manly), pp. 225-38. University of Otago Press, Dunedin.

Manly, B.F.J. (1997a). *Randomization, Bootstrap and Monte Carlo Methods in Biology*, 2nd edit. Chapman and Hall, London.

Manly, B.F.J. (1997b). *RT, A Program for Randomization Testing, Version 2.1*. Centre for Applications of Statistics and Mathematics, University of Otago, P.O. Box 56, Dunedin.

Manly, B.F.J. (2000). Statistics in the new millenium: some personal views. *Proceedings of the 11th Conference on Applied Statistics in Agriculture*, Kansas State University, Manhattan, 1999 (Ed. G.A. Milliken), pp. 1-13. Department of Statistics, Kansas State University.

Manly, B.F.J. and MacKenzie, D. (2000). A cumulative sum type of method for environmental monitoring. *Environmetrics* 11: 151-66.

Manly, B.F.J. and Walshe, K. (1999). The development of the population management plan for the New Zealand sea lion. In *Marine Mammal Survey and Assessment Methods* (Eds. G.W. Garner, S.C. Amstrup, J.L. Laake, B.F.J. Manly, L.L. McDonald and D.G. Robertson), pp. 271-83. Balkema, Rotterdam.

Mantel, N. (1967). The detection of disease clustering and a generalized regression approach. *Cancer Research* 27: 209-20.

Marr, J.C.A., Bergman, H.L., Lipton, J. and Hogstrand, C. (1995). Differences in relative sensitivity of naive and metals-acclimated brown and rainbow trout exposed to metals representative of the Clark Fork River, Montana. *Canadian Journal of Fisheries and Aquatic Science* 52: 2016-30.

McBride, G.B. (1999). Equivalence tests can enhance environmental science and management. *Australian and New Zealand Journal of Statistics* 41: 19-29.

McBride, G.B., Loftis, J.C. and Adkins, N.C. (1993). What do significance tests really tell us about the environment? *Environmental Management* 17: 423-32.

McCullagh, P. and Nelder, J.A. (1989). *Generalized Linear Models*. Chapman and Hall, London.

McDonald, L.L. and Erickson, W.P. (1994). Testing for bioequivalence in field studies: has a disturbed site been adequately reclaimed? In *Statistics in Ecology and Environmental Monitoring* (Eds. D.J. Fletcher and B.F.J. Manly), pp. 183-97. University of Otago Press, Dunedin.

McDonald, L.L., Erickson, W.P. and Strickland, M.D. (1995). Survey design, statistical analysis, and basis for statistical inferences in coastal habitat assessment: *Exxon Valdez* oil spill. In *Exxon Valdez Oil Spill: Fate and Effects in Alaskan Waters* (Eds. P.G. Wells, J.N. Butler and J.S. Hughes), pp. 296-311. American Society for Testing and Materials, Philadelphia.

McDonald, L.L. and Manly, B.F.J. (1989). Calibration of biased sampling procedures. In *Estimation and Analysis of Insect Populations* (Eds. L. McDonald, B. Manly, J. Lockwood and J. Logan), pp. 467-83. Springer-Verlag, Berlin.

McIntyre, G.A. (1952). A method for unbiased selective sampling, using ranked sets. *Australian Journal of Agricultural Research* 3:385-90.

McQueen, S.M. and Lloyd, B.D. (2000). The impact of 1080 pest control operations on invertebrates at Rangataua Forest, North Island, New Zealand. In preparation.

Mead, R. (1974). A test for spatial pattern at several scales using data from a grid of contiguous quadrats. *Biometrics* 30: 295-307.

Mead, R. (1988). *The Design of Experiments: Statistical Principles for Practical Applications*. Cambridge University Press, Cambridge.

Mead, R., Curnow, R.N. and A.M. Hasted (1993). *Statistical Methods in Agriculture and Experimental Biology*, 2nd edit. Chapman and Hall, London.

Miller, M.P. (1998). MANTEL-STRUC, A Windows program for the detection of population structure via Mantel tests. Available from the web site herb.bio.nau.edu/~miller/mantel.htm.

Minitab Inc. (1994). *MINITAB Reference Manual, Release 10 for Windows*. Minitab Inc., State College, Pennsylvania 16801, USA.

Mitchell, T. (1997). Northeast Brazil rainfall anomaly index. Fortaleza data available at the web site tao.atmos.washington.edu/data_sets/brazil.

Mohn, E. and Volden, R. (1985). Acid precipitation: effects on small lake chemistry. In *Data Analysis in Real Life Environment: Ins and Outs of Solving Problems* (Eds. J.F. Marcotorchino, J.M. Proth and J. Janssen), pp. 191-6. Elsevier, Amsterdam.

Montgomery, D.C. (1991). *Introduction to Statistical Quality Control*, 2nd edit. Wiley, New York.

Muir, W.D., Smith, S.G., Iwamoto, R.N., Kamikawa, D.J., McIntyre, K.W., Hockersmith, E.E., Sandford, B.P., Ocker, P.A., Ruehle, T.E., Williams, J.G. and Skalski, J.R. (1995). *Survival Estimates for the Passage of Juvenile Chinook Salmon through Snake River Dams and Reservoirs: Annual Report 1994*. Bonneville Power Administration, P.O. Box 3621, Portland, Oregon.

NCSS Statistical Software (1995). *PASS 1.0, Power Analysis and Sample Size for DOS*. NCSS Statistical Software, 329 North 1000 East, Kaysville, Utah 84037.

Nelder, J.A. (1999). Statistics for the millenium: from statistics to statistical science. *Statistician* 48: 257-69.

Nelder, J.A. and Wedderburn, R.W.M. (1972). Generalized linear models. *Journal of the Royal Statistical Society* A135: 370-84.

Neter, J., Wasserman, W. and Kutner, M.H. (1983). *Applied Linear Regression Models*. Irwin, Homewood, Illinois.

Newman, M.C., Greene, K.D. and Dixon, P.M. (1995). *Uncensor, Version 4.0*. Savannah River Ecology Laboratory, Aiken, South Carolina 29801. Manual and computer program available from the web site www.vims.edu/env/departments/riskchem/software/vims_software.html.

Norton, D.A. and Ogden, J. (1987). Dendrochronology: a review with emphasis on New Zealand applications. *New Zealand Journal of Ecology* 10: 77-95.

Nychka, D., Piegorsch, W.W. and Cox, L.H., Eds. (1998). *Case Studies in Environmental Statistics*. Springer-Verlag, New York.

Oakes, M. (1986). *Statistical Significance: A Commentary for the Social and Behavioural Sciences*. Wiley, New York.

Osenberg, C.W., Sarnelle, O. and Goldberg, D.E. (1999). Special feature on meta-analysis on ecology: concepts, statistics, and applications. *Ecology* 80: 1103-4.

Osenberg, C.W., Schmitt, R.J., Holbrook, S.J., Abu-Saba, K.E. and Flegal, A.R. (1994). Detection of environmental impacts: natural variability, effect size, and power analysis. *Ecological Applications* 4: 16-30.

Overrein, L.N., Seip, H.M. and Tollan, A. (1980). *Acid Precipitation - Effects on Forest and Fish: Final Report*. Norwegian Institute for Water Research, Oslo.

Overton, W.S. and Stehman, S.V. (1995). Design implications of anticipated data uses for comprehensive environmental monitoring programmes. *Environmental and Ecological Statistics* 2: 287-303.

Overton, W.S. and Stehman, S.V. (1996). Desirable design characteristics for long-term monitoring of ecological variables. *Environmental and Ecological Statistics* 3: 349-61.

Overton, W.S., White, D. and Stevens, D.L.. (1991). *Design Report for EMAP, the Environmental Monitoring and Assessment Program*. U.S. Environmental Protection Agency report EPA/600/3-91/053, Washington, DC.

Page, D.S., Gilfillan, E.S., Boehm, P.D. and Harner, E.J. (1995). Shoreline Ecology Program for Prince William Sound, Alaska, following the *Exxon Valdez* oil spill: Part 1- study design and methods. In *Exxon Valdez Oil Spill: Fate and Effects in Alaskan Waters* (Eds. P.G. Wells, J.N. Butler and J.S. Hughes), pp. 263-95. American Society for Testing and Materials, Philadelphia.

Page, E.S. (1955). A test for a change in a parameter occurring at an unknown time point. *Biometrika* 42: 523-6.

Page, E.S. (1961). Cumulative sum control charts. *Technometrics* 3: 1-9.

Palisade Corporation (2000). @Risk 4.0 spreadsheet add-on for Monte Carlo simulation. Information available from the web site www.palisade.com/.

Pannatier, Y. (1996). *VARIOWIN Software for Spatial Data in 2D*. Springer, New York.

Patil, G.P. (1995). Composite sampling. *Environmental and Ecological Statistics* 2: 169-79.

Patil, G.P. and Rao, C.R., Eds. (1994). *Handbook of Statistics 12: Environmental Statistics*. North-Holland, Amsterdam.

Patil, G.P., Sinha, A.K. and Taille, C. (1994). Ranked set sampling. In *Handbook of Statistics 12: Environmental Statistics* (Eds. G.P. Patil and C.R. Rao), pp. 167-200. North-Holland, Amsterdam.

Peres-Neto, P.R. (1999). How many tests are too many? The problem of conducting multiple ecological inferences revisited. *Marine Ecology Progress Series* 176: 303-6.

Perry, J.N. (1995a). Spatial aspects of animal and plant distribution in patchy farmland habitats. In *Ecology and Integrated Farming Systems* (Eds. D.M. Glen, M.P. Greaves and H.M. Anderson), pp. 221-42. Wiley, London.

Perry, J.N. (1995b). Spatial analysis by distance indices. *Journal of Animal Ecology* 64: 303-14.

Perry, J.N. (1998). Measures of spatial pattern and spatial association for counts of insects. In *Population and Community Ecology for Insect Management and Conservation* (eds. J. Baumgärtner, P. Bradmayr and B.F.J. Manly), pp. 21-33. Balkema, Rotterdam.

Perry, J.N. (2000). *SADIE: Spatial Analysis by Distance Indices*. Available from the IACR-Rothamstead web site at www.res.bbsrc.ac.uk/entnem/about/projects/joeperry/joeperry.htm.

Perry, J.N. and Hewitt, M. (1991). A new index of aggregation for animal counts. *Biometrics* 47: 1505-18.

Peterman, R.M. (1990). Statistical power analysis can improve fisheries research and management. *Canadian Journal of Fisheries and Aquatic Science* 47: 2-15.

Rasmussen, P.W., Heisey, D.M., Nordheim, E.V. and Frost, T.M. (1993). Time series intervention analysis: unreplicated large-scale experiments. In *Design and Analysis of Ecological Experiments* (Eds. S.M. Scheiner and J. Gurevitch), pp. 138-58. Chapman and Hall, New York.

Rice, W.R. (1990). A consensus combined p-value test and the family-wide significance of component tests. *Biometrics* 46: 303-8.

Ripley, B.D. (1981). *Spatial Statistics*. Wiley, New York.

Rowan, D.J., Rasmussen, J.B. and Kalff, J. (1995). Optimal allocation of sampling effort in lake sediment studies. *Canadian Journal of Fisheries and Aquatic Sciences* 52: 2146-58.

Ross, N.P. and Stokes, L. (1999). Editorial: special issue on statistical design and analysis with ranked set samples. *Environmental and Ecological Statistics* 6: 5-9.

Särkkä, A. (1993). Pseudo-likelihood approach for pair potential estimation of Gibbs processes. *Jyväskylä Studies in Computer Science, Economics and Statistics* 22. University of Jyväskylä, Finland.

Saville, D.J. (1986). The inconsistency of multiple comparison procedures. In *Pacific Statistical Congress* (Eds. I.S. Francis, B.F.J. Manly and F.C. Lam), pp. 286-9. Elsevier, Amsterdam.

Saville, D.J. (1990). Multiple comparison procedures: the practical solution. *American Statistician* 44: 174-80.

Scheaffer, R.L., Mendenhall, W. and Ott, L. (1990). *Elementary Survey Sampling*, 4th Edit. PWS-Kent, Boston.

Schipper, M. and Meelis, E. (1997). Sequential analysis of environmental monitoring data: refining SPRT's for testing against a minimal relevant trend. *Journal of Agricultural, Biological and Environmental Statistics* 2: 467-89.

Schmoyer, R.L., Beauchamp, J.J., Brandt, C.C. and Hoffman, F.O. (1996). Difficulties with the lognormal model in mean estimation and testing. *Environmental and Ecological Statistics* 3: 81-97.

Schneider, H. (1986). *Truncated and Censored Samples from Normal Populations*. Marcel Dekker, New York.

Schuirmann, D.J. (1987). A comparison of the two one-sided tests procedure and the power approach for assessing the equivalence of average bioavailability. *Journal of Pharmacokinetics and Biopharmaceutics* 15: 657-80.

Sherley, G. and Wakelin, M. (1999). Impact of monofluroacetate ("1080") on forest invertebrates at Ohakune, North Island, New Zealand. *New Zealand Journal of Zoology* (in press).

Shewhart, W.A. (1931). *Economic Control of Quality of Manufactured Product*. Van Nostrand, New York.

Siegel, S. (1956). *Nonparametric Statistics for the Behavioural Sciences*. McGraw-Hill, New York.

Skalski, J.R. (1990). A design for long-term status and trends. *Journal of Environmental Management* 30: 139-44.

Skalski, J.R., and Robson, D.S. (1992). *Techniques for Wildlife Investigations: Design and Analysis of Capture Data*. Academic Press, San Diego.

Smith, R.L. (1993). Long-range dependence and global warming. In *Statistics for the Environment* (Eds. V. Barnett and K.F. Turkman), pp. 141-61.

Smith, V.H. and Shapiro, J. (1981). Chlorophyll-phosphorus relations in individual lakes. Their importance in relation to lake restoration strategies. *Environmental Science and Technology* 15: 444-51.

SPSS Inc. (1997). SPSS for Windows, Release 8. SPSS Inc., 233 S. Wacker Drive, Chicago, Illinois 60606, USA.

Steel, R.G.D. and Torrie, J.H. (1980). *Principles and Procedures of Statistics: A Biometrical Approach*. McGraw-Hill, New York.

Stehman, S.V. and Overton, W.S. (1994). Environmental sampling and monitoring. In *Handbook of Statistics 12: Environmental Statistics* (Eds. G.P. Patil and C.R. Rao), pp. 263-306. Elsevier, Amsterdam.

Stevens, D.R. and Olsen, A.R. (1991). Statistical issues in environmental monitoring and assessment. *Proceedings of the Section on Statistics and the Environment*, pp. 76-85. American Statistical Association, Alexandria, Virginia.

Stewart-Oaten, A., Murdoch, W.W. and Parker, K.R. (1986). Environmental impact assessment: "pseudoreplication" in time? *Ecology* 67: 929-40.

Sullivan, J.H. and Woodhall, W.H. (1996). A control chart for preliminary analysis of individual observations. *Journal of Quality Technology* 28: 265-78.

Sunspot Index Data Center (1999). *Yearly Sunspot Data*. The Royal Observatory of Belgium website at www.oma.be/KSB-ORB/SIDC/.

Swed, F.S. and Eisenhart, C. (1943). Tables for testing randomness of grouping in a sequence of alternatives. *Annals of Mathematical Statistics* 14: 83-6.

Taylor, C.H. and Loftis, J.C. (1989). Testing for trend in lake and ground water quality time series. *Water Resources Bulletin* 25: 715-26.

Thompson, S.K. (1992). *Sampling*. Wiley, New York.

Troendle, J.F. and Legler, J.M. (1998). A comparison of one-sided methods to identify significant individual outcomes in a multiple outcome setting: stepwise tests or global tests with closed testing. *Statistics in Medicine* 17: 1245-60.

Underwood, A.J. (1994). On beyond BACI: sampling designs that might reliably detect environmental disturbances. *Ecological Applications* 4: 3-15.

Underwood, A.J. (1997). *Experiments in Ecology: Their Logical Design and Interpretation Using Analysis of Variance*. Cambridge University Press, Cambridge.

United States Environmental Protection Agency (1989a). *Methods for Evaluating the Attainment of Cleanup Standards. Volume 1: Soils and Solid Media*. EPA Report 230/02-89-042, Office of Policy, Planning and Evaluation, Washington DC.

United States Environmental Protection Agency (1989b). *Record of Decision for Koppers Superfund Site, Oroville, California*. United States Environmental Protection Agency, Region IX, San Francisco, California.

United States Environmental Protection Agency (1994). *Guidance for the Data Quality Objectives Process*. Report EPA/600/R-96/055, Office of Research and Development, Washington, DC.

United States Environmental Protection Agency (1997). *Guiding Principles for Monte Carlo Analysis*. Report EPA/630/R-97/001, US Environmental Protection Agency, Washington, DC.

United States Environmental Protection Agency (1998). *Guidance for Data Quality Assessment: Practical Methods for Data Analysis*. Report EPA/600/R-96/084, Office of Research and Development, Washington, DC.

United States Environmental Protection Agency (1999). *Policy for Use of Probabilistic Analysis in Risk Assessment at the US EPA*. Web page www.epa.gov/ncea/mcpolicy.htm.

United States Environmental Protection Agency (2000). *Risk Assessment Forum*. Web page www.epa.giv/ncea/raf/index.ttml.

United States Environmental Protection Agency and United States Army Corps of Engineers (1998). *Evaluation of Dredged Material Proposed for Discharge in Water of the U.S. - Testing Manual*. Report EPA-823-B-98-004, Washington, DC.

United States Office of Environmental Management (1997). Data Quality Objectives: why use the DQO process? Web page etd.pnl.gov:2080/DQO/why.html.

Urquhart, N.S., Overton, W.S. and Birkes, D.S. (1993). Comparing sampling designs for monitoring ecological status and trends: impact of temporal patterns. In *Statistics for the Environment* (Eds. V. Barnett and K.F. Turkman), pp. 71-85. Academic Press, Chichester.

Von Ende, C.N. (1993). Repeated-measures analysis: growth and other time-dependent measures. In *Design and Analysis of Ecological Experiments* (Eds. S.M. Scheiner and J. Gurevitch), pp. 113-58. Chapman and Hall, New York.

Wells, P.G., Butler, J.N. and Hughes, J.S. (1995). Introduction, overview, issues. In *Exxon Valdez Oil Spill: Fate and Effects in Alaskan Waters* (Eds. P.G. Wells, J.N. Butler and J.S. Hughes), pp. 3-38. American Society for Testing and Materials, Philadelphia.

Westfall, P.H. and Young, S.S. (1993). *Resampling-Based Multiple Testing: Examples and Methods for p-Value Adjustment*. Wiley, New York.

Westlake, W. J. (1988). Bioavailability and bioequivalence of pharmaceutical formulations. In *Biopharmaceutical Statistics for Drug Development* (Ed. K. E. Peace), pp. 329-52. Marcel Dekker, New York.

Wiens, B.L. (1999). When lognormal and gamma models give different results: a case study. *American Statistician* 53: 89-93.

Winston, W.L. (1996). *Simulation Modelling Using @RISK*. Duxbury Press, Belmont.

Wolter, K.M. (1984). An investigation of some estimators of variance for systematic sampling. *Journal of the American Statistical Association* 79: 781-90.

Yates, S.R. and Yates, M.V. (1990). Geostatistics for Waste Management: User's Manual for the GEOPACK (Version 1.0) Geostatistical Software System. Document EPA/600/8-90/004, Office of Research and Development, United States Environmental Protection Agency, ADA, Oklahoma, USA. Available from the web site www.epa.gov/ahaazvuc/geopack.html.

Younger, M.S. (1985). *A First Course in Linear Regression*. Duxbury Press, Boston.

Zetterqvist, L. (1991). Statistical estimation and interpretation of trends in water quality time series. *Water Resources Research* 27: 1637-48.

Author Index

Subject Index